McGraw-Hill Mathematics

Transition Handbook

Bridge the Gaps!

- What Do I Need to Know?
- Skill Builder
- Challenge

Teacher Guide 1

McGraw-Hill School Division

New York   Farmington

McGraw-Hill School Division

A Division of The McGraw-Hill Companies

Copyright © McGraw-Hill School Division,
a Division of the Educational and Professional Publishing Group of The McGraw-Hill Companies, Inc.
All rights reserved. Permission granted to reproduce for use with McGraw-Hill MATHEMATICS.

McGraw-Hill School Division
Two Penn Plaza
New York, New York 10121-2298

Printed in the United States of America

ISBN 0-02-100203-7 / 1

1 2 3 4 5 6 7 8 9 066 05 04 03 02 01 00

GRADE 1 Contents

To the Teacher viii

Chapter 1
Numbers to 20

Inventory Test Blackline Masters 1A
Inventory Test Prerequisite Skills Chart 1C
More or Fewer 2
Copy a Pattern 4
Extend a Pattern 5
Numbers to 5 6
Before, After, and Between 8
Challenge: Patterns 10
Challenge: Race Around the Pond 12

Chapter 2
Addition Concepts

Inventory Test Blackline Masters 13A
Inventory Test Prerequisite Skills Chart 13C
Numbers to 8 14
Compare Numbers 16
Concept of Addition 18
Same Numbers 20
Add Sums to 5 22
Challenge: Addition Town 24
Challenge: Addition Dominoes 26

Chapter 3
Addition Strategies and Facts to 12

Inventory Test Blackline Masters	27A
Inventory Test Prerequisite Skills Chart	27C
Numbers to 12	28
Order Numbers to 12	30
Add Sums to 10	32
Patterns	34
Challenge: Greatest Sum Game	36
Challenge: Addition Concentration	38

Chapter 4
Subtraction Concepts

Inventory Test Blackline Masters	39A
Inventory Test Prerequisite Skills Chart	39C
Numbers to 8	40
More or Fewer	41
Add Sums to 8	42
Same Number	44
Add with Zero	46
Challenge: Addition Dot-to-Dot	48
Challenge: Number Shapes	49

Chapter 5
Subtraction Strategies and Facts to 12

Inventory Test Blackline Masters	49A
Inventory Test Prerequisite Skills Chart	49C
Numbers to 12	50
Order Numbers to 12	52
Add Sums to 12	54
Concept of Subtraction	56
Subtract from 8	58
Challenge: Parking Lot Count	60
Challenge: Add and Subtract in Circles	61

Chapter 6
Data and Graphs

Inventory Test Blackline Masters	61A
Inventory Test Prerequisite Skills Chart	61C
Numbers to 20	62
More or Fewer	64
Add Sums to 12	66
Subtract from 8	68
Challenge: Classroom Count	70
Challenge: A Different Maze	71

Chapter 7
Place Value and Patterns

Inventory Test Blackline Masters	71A
Inventory Test Prerequisite Skills Chart	71C
Numbers to 20	72
Same Number	74
More or Fewer	75
Order Numbers to 20	76
Patterns	77
Compare Numbers to 20	78
Challenge: Go Fish for Numbers	80
Challenge: Secret Number Code	81

Chapter 8
Money

Inventory Test Blackline Masters	81A
Inventory Test Prerequisite Skills Chart	81C
Skip Count by 5s	82
Skip Count by 10s	83
Number Patterns: 1 More	84
Number Patterns: 10 More	85
Compare Numbers to 100	86
Add Sums to 12	88
Subtract from 12	90
Challenge: Toy Store Game	92
Challenge: Addition Search/Toy Store Sale	94

Chapter 9
Addition and Subtraction Strategies and Facts to 20

Inventory Test Blackline Masters	95A
Inventory Test Prerequisite Skills Chart	95C
Numbers to 20	96
Patterns	97
Add Sums to 12: Concept of Addtion	98
Add Sums to 12: Counting On	99
Add Sums to 12: Doubles	100
Add Sums to 12: Doubles + 1	101
Subtract from 12: Concept of Subtraction	102
Subtract from 12: Count Back to Subtract	104
Subtract from 12: Doubles	105
Related Addition Facts	106
Related Subtraction Facts	108
Missing Addends	110
Challenge: Go Shopping	112
Challenge: All Around the Mall	114

Chapter 10
Time

Inventory Test Blackline Masters	115A
Inventory Test Prerequisite Skills Chart	115C
Numbers to 100	116
Same Number	118
Before and After	119
Ordinals	120
Order Numbers to 100	122
Estimate Time	124
Challenge: Mystery Numbers	126
Challenge: Picturing Time	127

Chapter 11
Measurement

Inventory Test Blackline Masters	127A
Inventory Test Prerequisite Skills Chart	127C
Numbers to 100	128
Longer or Shorter	130
More Than, Less Than, About the Same As	132
More Than, Less Than, About the Same As	134
Challenge: The Long and Short of It	136
Challenge: Heavy Duty	137

Chapter 12
Geometry

Inventory Test Blackline Masters.....................137A
Inventory Test Prerequisite Skills Chart...............137C
Inside, Outside.......................................138
Follow Directions 140
Sort by Shape: Rolls..................................142
Sort by Shape: Same Shape.............................144
Sort by Size..146
Challenge: The Same Shape Game........................148
Challenge: Same-Size Concentration/Shape
Detective...150

Chapter 13
Fractions

Inventory Test Blackline Masters.....................151A
Inventory Test Prerequisite Skills Chart...............151C
Sort by Size..152
Equal Number 154
Numbers to 10...156
Challenge: Design a Beach Blanket.....................158
Challenge: Domino Fun.................................159

Chapter 14
Add and Subtract 2-Digit Numbers

Inventory Test Blackline Masters.....................159A
Inventory Test Prerequisite Skills Chart...............159C
Add Sums to 20..160
Add Sums to 20: Counting On 161
Add Sums to 20: Doubles...............................162
Add Sums to 20: Doubles + 1...........................163
Count to 100..164
Names for Numbers 166
Subtract from 20......................................168
Subtract from 20: Count Back 169
Subtract Doubles......................................170
Subtract Related Facts 171
Challenge: Number Names...............................172
Challenge: Subtraction Match 173

To the Teacher

Welcome to *McGraw-Hill Mathematics Transition Handbook: Bridge the Gaps!* The goal of these materials is to provide assessment and instruction in the prerequisite skills that some of your students need to be successful in math at this grade level.

For each chapter of the *McGraw-Hill Mathematics* student text, there is a 2-page inventory test called *What Do I Need To Know?* You will find these inventory tests as blackline masters on the A and B pages in this Teacher Guide. The results of the tests will help you diagnose any gaps in student knowledge. You can then provide students with materials needed to reteach or challenge them as appropriate.

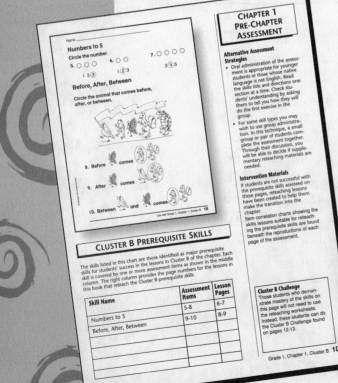

The charts found on the C and D pages following the blackline masters will prescribe a special *Skill Builder* lesson in the handbook for each test item that a student answers incorrectly.

viii

The *Skill Builder* lessons are presented in language that is simple and direct. The lessons are highly visual and have been designed to keep reading to a minimum.

The Learn section begins with a student asking *What Can I Do?* This section provides stepped-out models and one or more strategies to help bridge any gaps in the student's knowledge. Following this is *Try It*, a section of guided practice, and *Power Practice*, a section containing exercises to ensure that your students acquire the math power they need to be successful in each chapter of their mathematics textbook.

Two *Challenge* activities appear at the end of each chapter in the handbook. These provide a variety of math experiences for students who had no difficulty with the inventory test. Students will enjoy working on the puzzles, riddles, codes, and other more challenging formats. The *Challenge* activities will provide an opportunity for your more advanced students to work independently, allowing you to focus attention on those who need additional instruction before they work on the lessons in their math text.

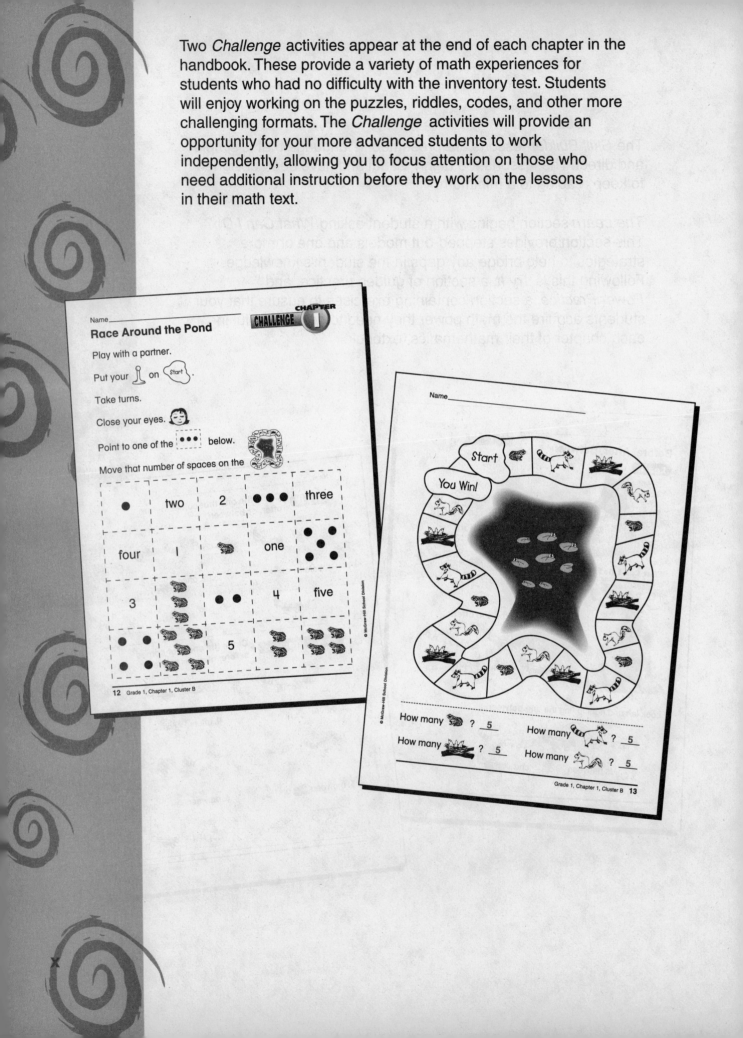

The Teacher Guide provides a complete lesson plan for each *Skill Builder* and *Challenge*. Each *Skill Builder* lesson plan includes a lesson objective, *Getting Started* activities, teaching suggestions, and questions to check the student's understanding. There is also a section called *What If the Student Can't*, which offers additional activities in case a student needs more support in mastering an essential prerequisite skill or lacks the understanding needed to complete the *Skill Builder* exercises successfully.

At least one *Skill Builder* lesson plan in each chapter has a feature called *Learn with Partners & Parents*. This activity is intended for students to use at home with parents or siblings or at school with a classmate-partner to practice a math skill in a game-like setting.

The lesson plan for each *Challenge* includes a lesson objective along with suggestions for introducing and using the *Challenge*.

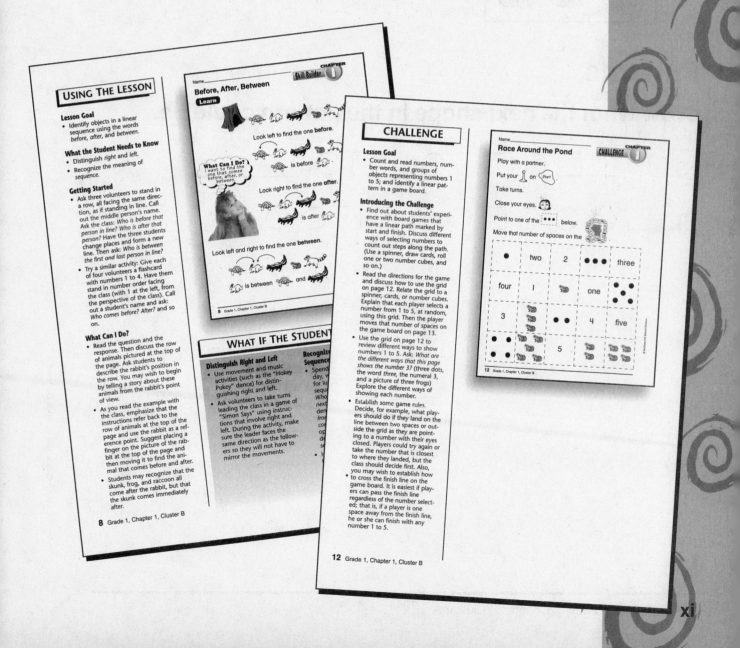

xi

Name _____

CHAPTER 1 ▶ What Do I Need To Know?

More or Fewer

Circle the one that has more.

1.

2.

Patterns

Draw what the next shape in the pattern could be.

3. ○ □ □ ○ □ □ ○ □ □ ○ _____

4. □ ○ ○ ○ □ ○ ○ ○ □ ○ ○ ○ _____

Name _____

Numbers to 5

Circle the number.

5. ○ ○ ○   6. ○ ○   7. ○ ○ ○ ○

   1 2 3   1 2 3   3 4 5

Before, After, Between

Circle the animal that comes before, after, or between.

8. Before  comes

9. After comes

10. Between and comes

Use with Grade 1, Chapter 1, Cluster B   **1B**

CHAPTER 1
PRE-CHAPTER ASSESSMENT

Assessment Goal
This two-page assessment covers skills identified as necessary for success in Chapter 1 Numbers to 20. The first page assesses the major prerequisite skills for Cluster A. The second page assesses the major prerequisite skills for Cluster B. When the Cluster A and Cluster B prerequisite skills overlap, the skill(s) will be covered in only one section.

Getting Started
- Allow students time to look over the two pages of the assessment. Point out the labels that identify the skills covered.
- Have students find math vocabulary terms used in the assessment. List vocabulary terms on the board as students identify them. If necessary, review the meanings of all essential math vocabulary.

Introducing the Assessment
- Explain to students that these pages will help you know if they are ready to start a new chapter in their math textbooks.
- Students who have transferred from another school may not have been introduced to some of these skills. Encourage students to do their best and assure them you will help them learn any needed skills.

Cluster A Challenge
Those students who demonstrate mastery of the skills on this page will not need to use the reteaching worksheets. Instead, these students can do the Cluster A Challenge found on pages 10-11.

Name _____

CHAPTER 1 — What Do I Need To Know?

More or Fewer
Circle the one that has more.

1.
2.

Patterns
Draw what the next shape in the pattern could be.

3. ○ □ □ ○ □ □ ○ ○ □ □ ○ ___
4. □ ○ ○ ○ □ ○ ○ ○ □ ○ ○ ___

1A Use with Grade 1, Chapter 1, Cluster A

CLUSTER A PREREQUISITE SKILLS

The skills listed in this chart are those identified as major prerequisite skills for students' success in the lessons in Cluster A of the chapter. Each skill is covered by one or more assessment items as shown in the middle column. The right column provides the page numbers for the lessons in this book that reteach the Cluster A prerequisite skills.

Skill Name	Assessment Items	Lesson Pages
More or Fewer	1-2	2-3
Patterns	3-4	4-5

Name _____

Numbers to 5

Circle the number.

5. ○ ○ ○   6. ○ ○   7. ○ ○ ○ ○
   1 2 ③   1 ② 3   3 ④ 5

Before, After, Between

Circle the animal that comes before, after, or between.

8. Before [butterfly] comes (frog)
9. After [butterfly] comes (squirrel)
10. Between [bird] and [squirrel] comes (frog)

Use with Grade 1, Chapter 1, Cluster B **1B**

CHAPTER 1 PRE-CHAPTER ASSESSMENT

Alternative Assessment Strategies
- Oral administration of the assessment is appropriate for younger students or those whose native language is not English. Read the skills title and directions one section at a time. Check students' understanding by asking them to tell you how they will do the first exercise in the group.
- For some skill types you may wish to use group administration. In this technique, a small group or pair of students complete the assessment together. Through their discussion, you will be able to decide if supplementary reteaching materials are needed.

Intervention Materials
If students are not successful with the prerequisite skills assessed on these pages, reteaching lessons have been created to help them make the transition into the chapter.
Item correlation charts showing the skills lessons suitable for reteaching the prerequisite skills are found beneath the reproductions of each page of the assessment.

CLUSTER B PREREQUISITE SKILLS

The skills listed in this chart are those identified as major prerequisite skills for students' success in the lessons in Cluster B of the chapter. Each skill is covered by one or more assessment items as shown in the middle column. The right column provides the page numbers for the lessons in this book that reteach the Cluster B prerequisite skills

Skill Name	Assessment Items	Lesson Pages
Numbers to 5	5-8	6-7
Before, After, Between	9-10	8-9

Cluster B Challenge
Those students who demonstrate mastery of the skills on this page will not need to use the reteaching worksheets. Instead, these students can do the Cluster B Challenge found on pages 12-13.

Grade 1, Chapter 1, Cluster B **1D**

USING THE LESSON

Lesson Goal
- Compare two groups of objects and identify which group has more objects or which group has fewer objects.

What the Student Needs to Know
- Count numbers 1 to 10.
- Recognize the concepts of *more* and *fewer*.

Getting Started
Place five green counters and four yellow counters on a table. Ask:
- *Which do we have more of, green or yellow?* (green)
- *How can you tell for sure?* (Encourage students to brainstorm ways to compare, including lining up the green and yellow counters in pairs or counting each group separately and then comparing the numbers.)
- *Which color has fewer counters?* (yellow)

What Can I Do?
- Read the question and the response. Then read and discuss the example. Ask:
- *How does matching each frog to a lily pad help show which group has more?* (You can see which group has some left over.)
- *How many frogs are there?* (3)
- *How many lily pads are there?* (4)
- *Are there more frogs or more lily pads?* (more lily pads)
- *Would it be possible for each frog to have its own lily pad? How can you tell?* (Yes, because there are more lily pads than frogs.)
- *If you had some yellow counters and some green counters, how could you tell which color you had more of?* (Match each yellow counter to a green counter.)

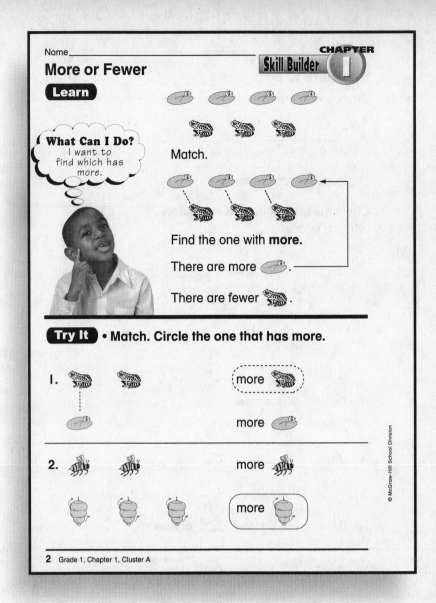

WHAT IF THE STUDENT CAN'T

Count Numbers 1 to 10
- Use number flashcards 1–10 and sets of objects, such as counters or connecting cubes, to reinforce the relationship between numbers and what they represent.
- Tell the student to do 10 jumping jacks and count each one aloud. Continue with other numbers up to 10.
- Have the student gather objects in specified, numbered groups. For example, tell him or her to make a group of six blocks. Count aloud as the blocks are placed in a group.

Recognize the Concepts of *More* and *Fewer*
Give the student an opportunity to practice relating more and fewer through a variety of physical or hands-on experiences.
- Allow the student to form two groups of objects—for example, placing a few crayons in one box and a few crayons in another box. Ask which box has more crayons.
- Have the student create two equal groups—for example, two boxes with the same number of crayons. Add another crayon to one box and ask which box has more.

Name_____

Power Practice

Match. Circle the one that has more.

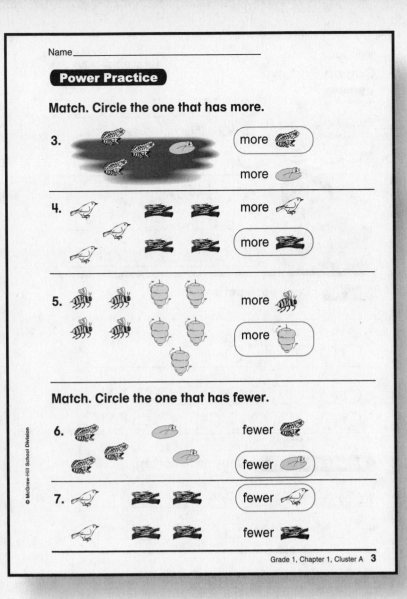

Match. Circle the one that has fewer.

Grade 1, Chapter 1, Cluster A **3**

USING THE LESSON

Try It
- Read the directions aloud and point out the sample answers. Check that students understand that the selections to the right of the groups of objects in each exercise are answer options.
- Relate exercise 1 to the example at the top of the page. Ask students to identify the two groups in the exercise. (frogs and lily pads)
- Discuss how to show the answer by circling the word and picture.
- For exercise 2, students might think of each bee flying to its own hive.
- For exercise 2, ask: *Which group has fewer, the group of bees or the group of hives?* (bees)

Power Practice
- Read the directions with the class. Point out that for exercises 3–5, students will look for the group that has more. For exercises 6 and 7, they will look for the group that has fewer.
- Review the exercises as a class. Ask volunteers to describe how they decided which group had more and which had fewer.

WHAT IF THE STUDENT CAN'T

Recognize the Concepts of *More* and *Fewer* (continued)

- Pair two students. Have them take a few steps across the room, matching each other's steps. Then ask one of them to take two more steps. Ask: *Who took fewer steps?*
- If a student is preoccupied with counting rather than comparing, use larger groups of about 15 and 25 similar objects (such as same-size marbles) so the quantity differential is obvious.
- If the student confuses size with quantity, have students compare groups of connecting cubes.

Complete the Power Practice
- Discuss each incorrect answer. Have the student model matching items from the two groups, identifying which group has items left over after matching, and selecting the answer.

Grade 1, Chapter 1, Cluster A **3**

USING THE LESSON

Lesson Goal
- Identify and duplicate a linear geometric pattern by drawing it.

What the Student Needs to Know
- Distinguish plane shapes (triangle, square, and circle).
- Recognize a linear pattern.

Getting Started
Find out what students know about patterns. Use red and blue counters (or a similar arrangement) to make an alternating pattern. Ask:
- *How many colors make up this pattern? What are they?* (two, red and blue)
- *How would you describe the pattern?* (red, blue, red, blue, and so on)

What Can I Do?
Read the question and the response. Then read and discuss the example. Ask:
- *How many shapes make up a "chunk"?* (3)
- *What makes up a chunk in this pattern?* (circle, square, square)
- *If you wanted to tell someone how to make this pattern, how could you describe it?* (Possible answer: repeat the chunk.)
- *What is the difference between a chunk of shapes and a pattern?* (The chunk is repeated to make the pattern.)

Try It
Read the directions aloud. Ask:
- *What is the chunk in the first pattern?* (square, circle)
- *What is the chunk in the second pattern?* (circle, circle, circle, square)

Power Practice
- Have students complete the practice items. Then review the answers.

WHAT IF THE STUDENT CAN'T

Distinguish Plane Shapes
- Use attribute blocks to reinforce basic plane shapes, focusing on squares, triangles, and circles. You might have the student sort the blocks by shape.
- Review the attributes of squares and triangles. Have the student describe the shapes by the number of sides and corners.
- Invite the student to practice drawing circles, squares, and triangles. Provide objects he or she can trace to create these plane shapes.

Recognize a Linear Pattern
- Use a simple hand-clapping exercise to practice sound patterns. Establish the pattern (for example, hands clap together two times, then both hands clap on the desk two times, and so on). Invite the student to join in. Then ask her or him to describe the pattern.

Complete the Power Practice
- Discuss each incorrect answer. Have the student describe the pattern and identify the chunk that repeats to create it.
- Have the student identify the shape that begins the pattern and the shape that ends it.

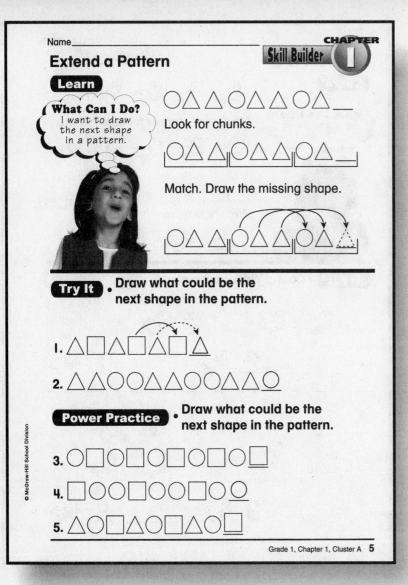

USING THE LESSON

Lesson Goal
- Continue a linear geometric pattern by drawing the next shape in the sequence.

What the Student Needs to Know
- Distinguish plane shapes (circle, triangle, rectangle, and square).
- Recognize a linear pattern.

Getting Started
Find out what students know about patterns. Have them use two different colors of counters to make a linear pattern (for example: green, red, red, green, red, red). Ask:
- *How would you describe this pattern?* (It's a repeat of green, red, and red.)
- *What is the chunk that repeats to form the pattern?* (green, red, red)
- *How could you continue the pattern?* (keep repeating the chunk)

What Can I Do?
Read the question and the response. Then read and discuss the example. Point out that the arrows are not part of the pattern. Ask:
- *In this pattern, which shape always comes after a circle?* (triangle)
- *What does the line under each group of shapes mean?* (It shows the chunk that repeats to make the pattern.)

Try It
Read the directions aloud. Have students identify the chunk in each of the patterns. Ask:
- *How does finding the chunk help you decide which shape to draw?* (You have to know the sequence that gets repeated to form the pattern.)

Power Practice
- Have students complete the practice items. Then review the answers.

WHAT IF THE STUDENT CAN'T

Distinguish Plane Shapes
- Have the student practice sorting attribute blocks by shape.
- Review the difference between squares and rectangles.
- Invite the student to practice drawing basic plane shapes.

Recognize a Linear Pattern
- Have the student verbalize the pattern in each exercise, saying aloud the shape names.
- Let pairs of students practice forming linear patterns with two colors of counters. Each student takes a turn making a pattern, with the partner copying it.

Complete the Power Practice
- Discuss each incorrect answer. Have the student re-create the pattern using attribute blocks. Shift the blocks to separate the chunks visually. Then ask the student to compare the partial chunk at the end with the whole chunk at the beginning.
- Draw the shapes that complete the chunk. Then have the student continue the pattern from that point.

USING THE LESSON

Lesson Goal
- Count objects and write the numeral that corresponds to the quantity of 1 to 5.

What the Student Needs to Know
- Recognize the concept of quantity.
- Read numerals 1 to 5.
- Write numerals 1 to 5.

Getting Started
With the class, play a guessing game for numbers 1 to 5. After each answer given, count aloud the items in the set. For example, ask:

- *How many fingers do you have on one hand?* (5)
- *How many eyes do you have?* (2)
- *How many ears?* (2)
- *How many heads?* (1)
- *How many limbs—that is, both legs and arms?* (4)

What Can I Do?
Read the question and the response. Then read and discuss the example. Ask:

- *In this example, what are we counting?* (butterflies)
- *What number is written on the line, and what does the number tell us?* (5. It's the number of butterflies in the example.)
- *If you covered the last butterfly with your hand, how many could you count?* (4) *What if you covered the last two?* (3)
- *If the butterflies were not in a straight row, would that change the number? How could you tell?* (No; you could count them.)

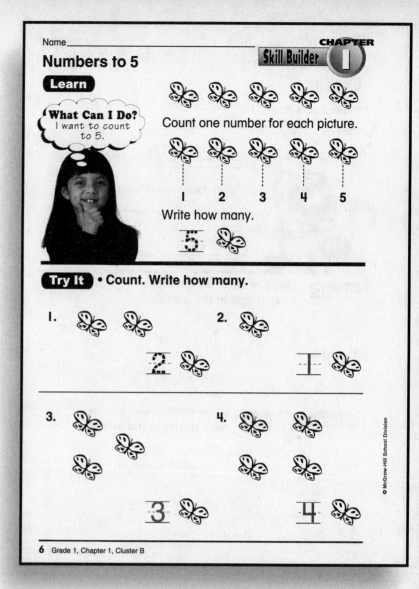

WHAT IF THE STUDENT CAN'T

Recognize the Concept of Quantity
- Provide a variety of number-counting picture books, or have students work together in small groups to make their own themed booklets for numbers 1–5. The first page would show one object, the second page would show two objects, and so on.
- Use number flashcards 1–5 and groups of objects, such as small plastic animals, to reinforce the relationship between numbers and what they represent. Arrange the objects in groups of 1 to 5, and ask the student to match a flashcard to the set of objects.

Read Numerals 1 to 5
- Pair students and have them practice using number flashcards 1–5. You may also wish to have the student practice reading number words.
- Challenge the student to find numbers in a variety of places: on a clock or calendar, on a computer keyboard, on a board game or playground hopscotch game, on package labels, and so on. Look for instances of 1 through 5.

Name_____

Power Practice • Count. Write how many.

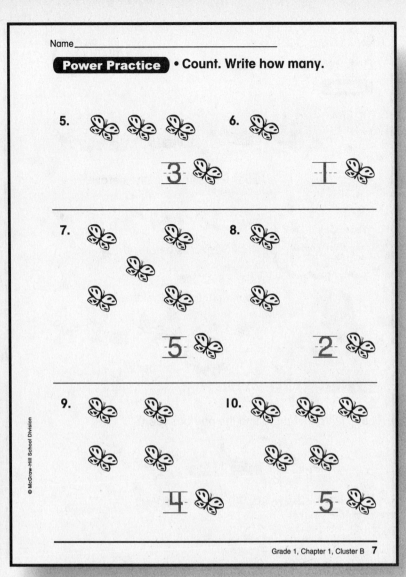

Grade 1, Chapter 1, Cluster B 7

WHAT IF THE STUDENT CAN'T

Write Numerals 1 to 5
- Have the student practice "air writing" as you verbalize the directions for writing each number in sequence. For example, for the number 2, say: *Start at the left and curve up and over to the right to make a half moon, then continue down to the left, past where you started, and draw a straight line across to the right.*
- Provide number cards that have direction arrows to show how to write the numbers. Allow the student to practice tracing the numbers with their fingers.

Complete the Power Practice
- Discuss each incorrect answer. Have the student model counting the butterflies and writing the appropriate number.

USING THE LESSON

Try It
- Read the directions aloud and point out the writing line on which to record the answer. Remind students to count only the butterflies for the exercise.
- Check that students understand they need to record only the total number of butterflies in the exercise; they will not be writing all of the numbers as they count.
- Encourage students to use the entire height of the writing line to write the number.
- Find out how students counted the butterflies in exercises 3 and 4. For example, they may begin at the top left and count in a clockwise direction. Suggest beginning the count from a different butterfly to see if students understand that it doesn't affect the result.

Power Practice
- Read the directions with the class, and make sure students understand which group of butterflies and which writing line go with each exercise.
- Invite volunteers to use five counters to reproduce the arrangement of butterflies in exercises 7 and 10. Talk about different ways to show five.
- Review *more* and *fewer* by comparing the groups of butterflies in exercises 5 and 6, 7 and 8, and 9 and 10. Have students find examples of same quantities as in exercises 7 and 10.

Learning with Partners and Parents
- Have students work in pairs to draw pictures that represent numbers 1 to 5 in a variety of ways. Collect all the pictures and have students sort them by number.
- Have partners list (or draw pictures to show) things that come in groups of 1 to 5. For example, two eyes or ears, four tires on a car, five points on a star.

Grade 1, Chapter 1, Cluster B **7**

USING THE LESSON

Lesson Goal
- Identify objects in a linear sequence using the words *before*, *after*, and *between*.

What the Student Needs to Know
- Distinguish *right* and *left*.
- Recognize the meaning of *sequence*.

Getting Started
- Ask three volunteers to stand in a row, all facing the same direction, as if standing in line. Call out the middle person's name. Ask the class: *Who is before that person in line? Who is after that person?* Have the three students change places and form a new line. Then ask: *Who is between the first and last person in line?*
- Try a similar activity: Give each of four volunteers a flashcard with numbers 1 to 4. Have them stand in number order facing the class (with 1 at the left, from the perspective of the class). Call out a student's name and ask: *Who comes before? After?* and so on.

What Can I Do?
- Read the question and the response. Then discuss the row of animals pictured at the top of the page. Ask students to describe the rabbit's position in the row. You may wish to begin by telling a story about these animals from the rabbit's point of view.
- As you read the example with the class, emphasize that the instructions refer back to the row of animals at the top of the page and use the rabbit as a reference point. Suggest placing a finger on the picture of the rabbit at the top of the page and then moving it to find the animal that comes before and after.
- Students may recognize that the skunk, frog, and raccoon all come after the rabbit, but that the skunk comes immediately after.

WHAT IF THE STUDENT CAN'T

Distinguish *Right* and *Left*
- Use movement and music activities (such as the "Hokey Pokey" dance) for distinguishing right and left.
- Ask volunteers to take turns leading the class in a game of "Simon Says" using instructions that involve right and left. During the activity, make sure the leader faces the same direction as the followers so they will not have to mirror the movements.

Recognize the Meaning of Sequence
- Spend a few minutes each day, when students line up for lunch or recess, to ask sequence questions such as: *Who is first in line? Who comes next?* Take turns naming students and asking: *Who is in front of that person? Who comes after?* Find similar opportunities to develop students' vocabulary with sequence.
- Help the student practice left-to-right sequencing in other ways, such as having them locate the first and last pages in a book and the first and last words on a page.

Name_____

Try It • Circle the one that comes before, after, or between.

1. before
2. after

Power Practice • Circle the one that comes before, after, or between.

3. before
4. after
5. between ___ and ___
6. between ___ and ___

Grade 1, Chapter 1, Cluster B 9

WHAT IF THE STUDENT CAN'T

Recognize the Meaning of Sequence (continued)

- Line up several colored blocks on the desk, and explain that the left-most block is the start of the row. Ask volunteers to arrange the blocks according to your instructions. For example, ask a student to put a blue block between two red blocks or to put a green block after a red block. As students become more adept, extend this to a logical reasoning activity: *A blue block comes after a green block. A red block is first in line. Which color is between blue and red?* (green)

Complete the Power Practice

- Discuss each incorrect answer. Have the student model how to locate the animal that comes before, after, or between. Have the student model selecting the appropriate response. If the student has difficulty, try numbering the animals from left to right. He or she may recognize more readily that 2 comes before 3, for example.

USING THE LESSON

Try It

- Read the directions aloud and review the sequence of animals in the row. Ask: *Which is the first animal in the row?* (frog) *The last?* (rabbit) *If you numbered the animals, which would be number 4?* (raccoon)
- Check that students understand that for exercises 1-2 they are referring to the row of animals in the Try It section (rather than the row at the top of page 8).
- Check that students understand how to read the exercises and select an appropriate answer by circling one of the two pictures for each exercise.

Power Practice

- Read the directions with the class and review the new row of animals. Make sure students understand that exercises 3–6 refer to this row.
- Invite students to discuss the strategies they used in locating the ones before and after. For example, with exercise 3, the likely strategy is to locate the rabbit in the row at the top of the Power Practice section and then move the finger to the animal at the left to find the raccoon. Then have students describe the strategies they used to find the one between. Ask: *Why might you need to use a different strategy?* (You need to locate two animals first.)
- To extend this activity, ask: *Which animals come between the frog and the skunk?* (raccoon, rabbit, and turtle). *Which two animals come before the rabbit?* (frog and raccoon).

Grade 1, Chapter 1, Cluster B 9

CHALLENGE

Lesson Goal
- Identify a linear geometric pattern hidden in a grid maze.

Introducing the Challenge
As appropriate, review basic plane shapes and the concepts of left and right and up and down. Read aloud the first sentence on the page. Have students identify the shapes that make up a chunk. Ask:

- *How can you tell what a chunk is?* (It's the group of objects that gets repeated. The lines under the four shapes show a chunk.)

- *Why is the order of the shapes important?* (It forms the pattern.)

Using the Challenge
- Read the directions. Check that students understand they will be searching the maze for the chunk of four shapes, not just a single shape at a time.

- Relate the first circle in the maze to the circle in the pattern above. Ask students to name the next three shapes in the pattern and to find them in the maze. Then ask them to look for the next circle that has a triangle after it. Remind students that they can move left, right, or down in the maze.

- Ask: *Why wouldn't you color the first triangle in the second row?* (A star comes after the triangle; it doesn't match the pattern.) Continue in this way until students have located the next chunk and colored the shapes.

- When students have completed the activity, ask them what was easy and what was difficult. Invite them to explain how they found the pattern in the maze.

- Some students may be ready to create their own pattern and maze. Have students work in small groups; allow them to exchange pattern mazes and give them a try.

10 Grade 1, Chapter 1, Cluster A

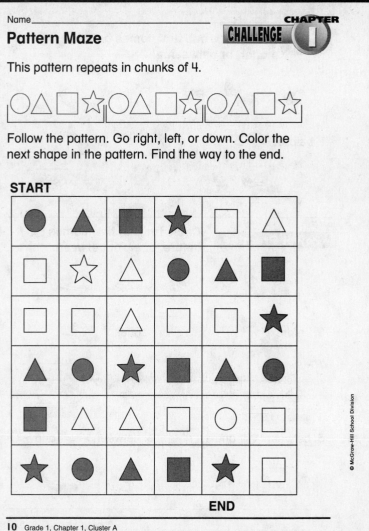

Name_____

Pattern Necklaces

CHALLENGE 1

Each 🪩 has a pattern. Find the pattern. Count the beads. Draw what could be the rest of the beads for each 🪩.

1. This 🪩 needs 10 beads in all.

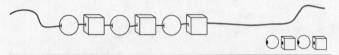

2. This 🪩 needs 9 beads in all.

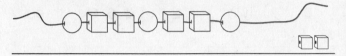

3. This 🪩 needs 10 beads in all.

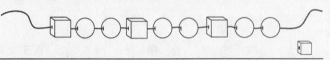

4. This 🪩 needs 9 beads in all.

Grade 1, Chapter 1, Cluster A   11

CHALLENGE

Lesson Goal
- Continue linear geometric patterns by drawing the next shapes in the sequence.

Introducing the Challenge
- Show a model of a sphere and cube and ask students to describe how they are different. Review counting to 10.
- Use large colored beads to introduce this activity. Ask students to choose three different colored beads to string. Suggest that you would like to make a necklace that repeats these three beads in the same order. Ask: *How many beads of each color will I need?* (3) After stringing the beads, count them from left to right. Then have the students say the pattern aloud. Talk about how the chunk of three colors is repeated three times.
- Give students an opportunity to create their own patterns using different colors of stringing beads.

Using the Challenge
- Read and discuss the directions. Then read the first exercise. Ask: *How many beads does this necklace have already?* (6) *How many more does it need so it has 10 in all?* (4) *How can you figure this out?* (Answers will vary; encourage brainstorming.) Provide manipulatives that students can use to duplicate the necklaces depicted if they need to.
- For each exercise, have students identify the chunk of shapes that repeat to form the pattern.
- Check that students understand how to record their answers by drawing the missing shapes in the correct order, to the right of the necklace. They can draw a square for each cube and a circle for each sphere. Have students count the beads in each necklace and then call out the shapes aloud.

Grade 1, Chapter 1, Cluster A   **11**

CHALLENGE

Lesson Goal
- Count and read numbers, number words, and groups of objects representing numbers 1 to 5; and identify a linear pattern in a game board.

Introducing the Challenge
- Find out about students' experience with board games that have a linear path marked by start and finish. Discuss different ways of selecting numbers to count out steps along the path. (Use a spinner, draw cards, roll one or two number cubes, and so on.)
- Read the directions for the game and discuss how to use the grid on page 12. Relate the grid to a spinner, cards, or number cubes. Explain that each player selects a number from 1 to 5, at random, using this grid. Then the player moves that number of spaces on the game board on page 13.
- Use the grid on page 12 to review different ways to show numbers 1 to 5. Ask: *What are the different ways that this page shows the number 3?* (three dots, the word *three*, the numeral 3, and a picture of three frogs) Explore the different ways of showing each number.
- Establish some game rules. Decide, for example, what players should do if they land on the line between two spaces or outside the grid as they are pointing to a number with their eyes closed. Players could try again or take the number that is closest to where they landed, but the class should decide first. Also, you may wish to establish how to cross the finish line on the game board. It is easiest if players can pass the finish line regardless of the number selected; that is, if a player is one space away from the finish line, he or she can finish with any number 1 to 5.

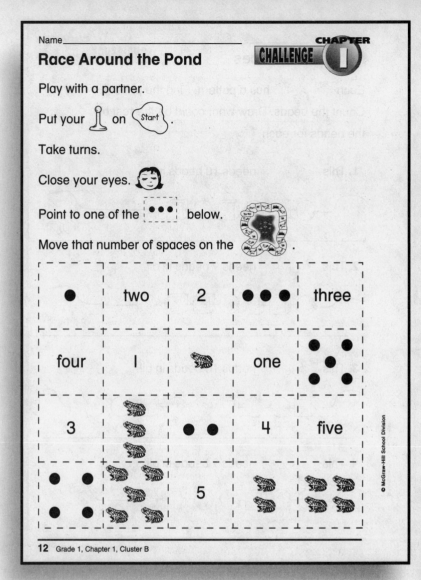

12 Grade 1, Chapter 1, Cluster B

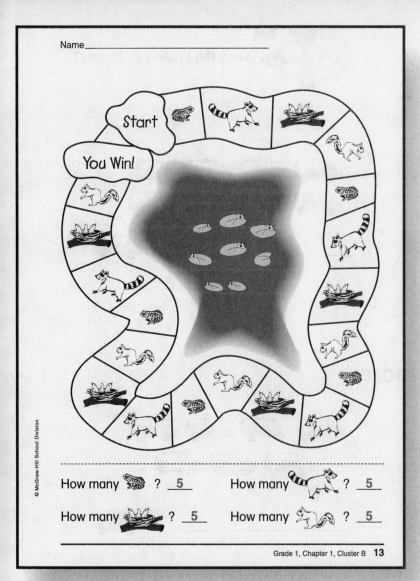

CHALLENGE

Using the Challenge

- Check that students understand how to play the game. Ask: *If you select the space that has a picture of five frogs, how many spaces do you move?* (5) Make sure students recognize that the picture means that they move five consecutive steps, not five steps that have frogs on them. Ask: *How do you know when you've won?* (when you cross the finish line)

- Have students run a finger along the game board path, noting the direction of play. Ask a volunteer to describe the pattern. The student should identify the items that form a chunk in the pattern. (frog, raccoon, nest, squirrel) Ask: *How many times is the chunk repeated?* (5) *Can you tell, without counting, how many pictures of bird's nests there are?* (5) *How do you know?* (The chunk repeats 5 times.)

- Use the game board to review the concepts of *before*, *after*, and *between*. Remind students that the path is curved like a circle; therefore, they can't use left and right to signal proximity to start or finish.

- Discuss the four questions beneath the game board. Check that students understand that these questions refer to the pattern on the game board, not to the number of times they land on those pictures as they play. Ask: *How did you find the total number of frogs?* (Likely response is, "By counting.")

- After students are familiar with the game, suggest that they keep track of how many moves it takes them to finish.

- Have students work in small groups to make their own game boards, using inked rubber stamps to create the pattern.

Grade 1, Chapter 1, Cluster B **13**

Name_____

Numbers to 8

CHAPTER 2 — What Do I Need To Know?

Write how many.

1.
_____

2.
_____

Compare Numbers

Write how many.
Circle the group that has more.

3.

_____   _____

Concept of Addition

Write how many in all.

4.

_____  in all.

5.

_____  in all.

13A Use with Grade 1, Chapter 2, Cluster A

Name _____

Same Numbers

**Write how many.
Circle the numbers that are the same.**

6.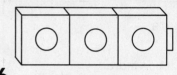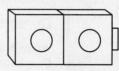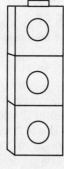

_____ _____ _____

7.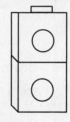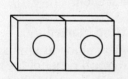

_____ _____ _____

Add Sums to 5

Add. Write each sum.

8.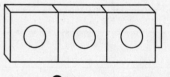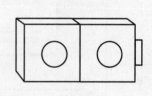

    3   +   2   = _____

9.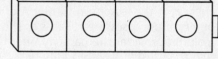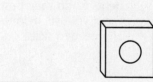

    4   +   1   = _____

10.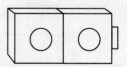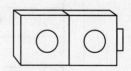

    2   +   2   = _____

CHAPTER 2 PRE-CHAPTER ASSESSMENT

Assessment Goal

This two-page assessment covers skills identified as necessary for success in Chapter 2 Addition Concepts. The first page assesses the major prerequisite skills for Cluster A. The second page assesses the major prerequisite skills for Cluster B. When the Cluster A and Cluster B prerequisite skills overlap, the skill(s) will be covered in only one section.

Getting Started

- Allow students time to look over the two pages of the assessment. Point out the labels that identify the skills covered.
- Have students find math vocabulary terms used in the assessment. List vocabulary terms on the board as students identify them. If necessary, review the meanings of all essential math vocabulary.

Introducing the Assessment

- Explain to students that these pages will help you know if they are ready to start a new chapter in their math textbooks.
- Students who have transferred from another school may not have been introduced to some of these skills. Encourage students to do their best and assure them you will help them learn any needed skills.

Cluster A Challenge

Those students who demonstrate mastery of the skills on this page will not need to use the reteaching worksheets. Instead, these students can do the Cluster A Challenge found on pages 24-25.

13C Grade 1, Chapter 2, Cluster A

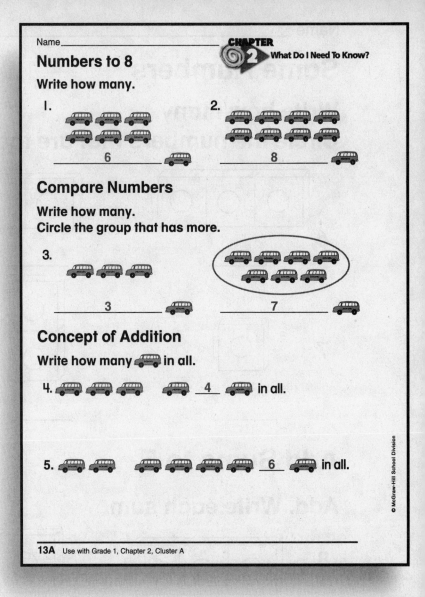

CLUSTER A PREREQUISITE SKILLS

The skills listed in this chart are those identified as major prerequisite skills for students' success in the lessons in Cluster A of the chapter. Each skill is covered by one or more assessment items as shown in the middle column. The right column provides the page number for the lessons in this book that reteach the Cluster A prerequisite skills.

Skill Name	Assessment Items	Lesson Pages
Numbers to 8	1-2	14-15
Compare Numbers	3	16-17
Concept of Addition	4-5	18-19

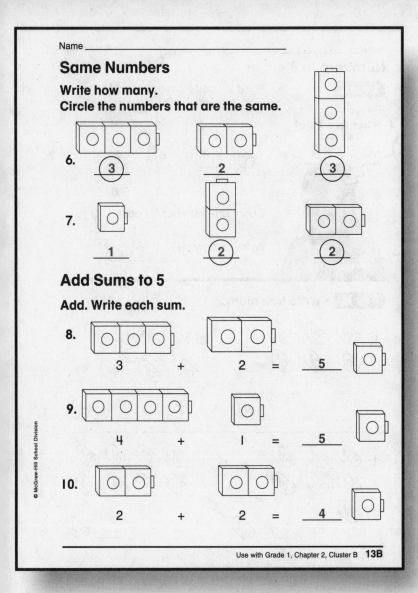

CHAPTER 2 PRE-CHAPTER ASSESSMENT

Alternative Assessment Strategies

- Oral administration of the assessment is appropriate for younger students or those whose native language is not English. Read the skills title and directions one section at a time. Check students' understanding by asking them to tell you how they will do the first exercise in the group.
- For some skill types you may wish to use group administration. In this technique, a small group or pair of students complete the assessment together. Through their discussion, you will be able to decide if supplementary reteaching materials are needed.

Intervention Materials

If students are not successful with the prerequisite skills assessed on these pages, reteaching lessons have been created to help them make the transition into the chapter.

Item correlation charts showing the skills lessons suitable for reteaching the prerequisite skills are found beneath the reproductions of each page of the assessment.

CLUSTER B PREREQUISITE SKILLS

The skills listed in this chart are those identified as major prerequisite skills for students' success in the lessons in Cluster B of the chapter. Each skill is covered by one or more assessment items as shown in the middle column. The right column provides the page numbers for the lessons in this book that reteach the Cluster B prerequisite skills

Skill Name	Assessment Items	Lesson Pages
Same Numbers	6-7	20-21
Add Sums to 5	8-10	22-23

Cluster B Challenge

Those students who demonstrate mastery of the skills on this page will not need to use the reteaching worksheets. Instead, these students can do the Cluster B Challenge found on pages 26-27.

Grade 1, Chapter 2, Cluster B **13D**

USING THE LESSON

Lesson Goal
- Count objects and write the numeral that corresponds to the quantity of 1 to 8.

What the Student Needs to Know
- Recognize the concept of quantity.
- Read numerals 1 to 8.
- Write numerals 1 to 8.

Getting Started
Pair students and have them play a counting game, such as ring toss or dropping pennies into a jar, using 8 items. Have them keep score with tally marks, noting how many of the 8 items hit the target. Compare scores. Ask:
- Who scored more than 3?
- Who scored more than 5?
- What was your highest score?
- How did you keep score?

What Can I Do?
Read the question and the response. Then read and discuss the example. Point out the numeral 8 on the writing line. Ask:
- What does it mean to "count one number for each one"? (Answers will vary. It means that you say a different number, in order, for each thing you count.)
- Are we counting these mailboxes across the page or down? (across)
- If we counted them in a different direction, would the final number be the same? (yes)
- Why does the writing line show the number 8? (That is how many mailboxes there are.)
- If you covered up the last mailbox, how many would you see? (7) What if you covered the last two? (6)

WHAT IF THE STUDENT CAN'T

Recognize the Concept of Quantity
- Have students work in pairs, using number flash cards 1–8 and sets of objects, such as counters, to reinforce the relationship between numbers and what they represent. Partners can arrange objects in different patterns to show the same number displayed.
- Provide opportunities for students to practice counting to 8 every day. Ask them to count people in line, pictures on a bulletin board, bicycles in racks, and so on.

Read Numerals 1 to 8
- Use number flash cards to reinforce reading numerals.
- Have on hand several picture books for counting, and allow students to read on their own.

Name_____

Power Practice • Write how many.

5. _5_
6. _7_
7. _6_
8. _4_
9. _8_
10. _6_
11. _7_
12. _3_

Grade 1, Chapter 2, Cluster A 15

WHAT IF THE STUDENT CAN'T

Write Numerals 1 to 8
- As a class, practice air writing numbers. Together, recite directions for writing each number.
- Make a set of number cards that have a variety of textures (sandpaper, velvet, fleece, and so on), and have students use the cards to practice writing, using a finger to show the direction of writing along the surface of the textured number.

Complete the Power Practice
- Discuss each incorrect answer. Have the student model counting and writing numbers.
- If students have difficulty keeping track of the pictures they count, suggest they put a check mark next to each one as they count it.

USING THE LESSON

Try It
- Read the directions aloud and point out the sample answer for exercise 1. Relate this exercise to the example at the top of the page and ask: *What does the example show that the exercise does not show?* (the numbers for counting each picture)
- Check that students understand that for each exercise they need only write the total number of mailboxes; they do not need to write the numbers as they count.
- Discuss the pattern created by each set of pictures. Compare similar sets (for example, those in exercises 1 and 4). Ask volunteers to describe the patterns. Discuss which numbers of objects, or which patterns of objects, are easily recognizable without counting. For example, students may recognize at a glance that exercise 4 has 8 mailboxes, but they may need to count the mailboxes in exercise 2.

Power Practice
- Ask students to describe their strategies for finding the total number of mailboxes in a set. Find out if they counted the mailboxes in each set or figured out the number based on the pattern.
- Ask students to compare exercises 9 and 11. Ask: *If you know how many mailboxes are in exercise 9, how might you use this information to help you with exercise 11?* (It's one less.)
- Discuss how different patterns can be used to show numbers. Ask: *How many different ways can you show six objects? Eight objects?* (Answers will vary.)

Grade 1, Chapter 2, Cluster A 15

USING THE LESSON

Lesson Goal
- Compare quantities for two groups of 1–8 objects.

What the Student Needs to Know
- Count and write numbers 1 to 8.
- Recognize the concepts of *more* and *fewer*.
- Distinguish right and left.

Getting Started
Find out what students know about comparing quantities of objects in groups. Ask:
- *How could you find out if there are more boys than girls in our class?* (Possible answer: Match each boy and girl and see who is left over.)
- *What if you want to find out how many more girls there are than boys?* (Count the number of boys or girls that are left over.)

What Can I Do?
Read the question and the response. Then read and discuss the example. Point out that it shows two ways to compare groups of objects. Ask:
- *What are we comparing in this example?* (the number of buses above the line and the number of buses below the line)
- *How many groups of objects are we comparing?* (2) *How do you know?* (The line separates the buses into two groups.)
- *How can you compare the buses in the two groups without counting them?* (Match each bus in one group with a bus in the other group, and count how many are left over after all are matched.)
- *Which way do you think is easier: counting and comparing numbers, or matching to see how many are left over?* (Answers will vary.)

WHAT IF THE STUDENT CAN'T

Count and Write Numbers 1 to 8
- Pair students and have them use flash cards and sets of objects to reinforce counting skills. One student holds up a number, and the partner displays that number of objects, counting while gathering.
- Provide opportunities for students to play board games that involve counting. Pair students who have strong counting skills with those who need more support.

Recognize the Concepts of *More* and *Fewer*
- Have students sort objects, such as connecting cubes, and determine which of two groups (for example, blue or yellow) has more.
- Help the class make pictographs (for example, to record favorite vegetables, given a choice of four or five). Ask questions that compare two sets of data using the pictograph. Encourage students to use the terms *more* and *fewer* in their responses.

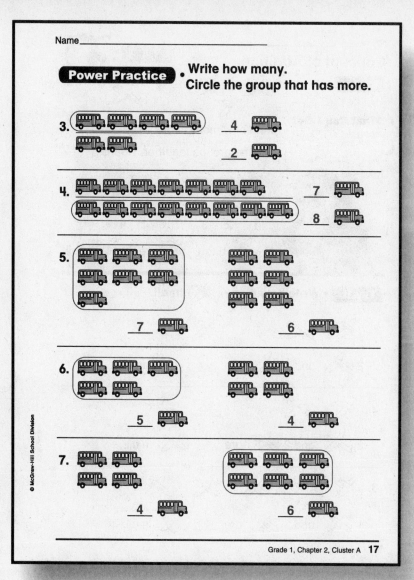

USING THE LESSON

Try It
Read and discuss the directions. Relate exercise 1 to the example. Point out that students may visually match the buses above the line with the first five buses below the line. Then they can see that there are still two buses left over below the line. Discuss how exercise 2 is presented differently. Check that students understand how to record their answers. Ask:

- *What are the two groups we are comparing in exercise 2?* (the group of buses on the left and the group of buses on the right)
- *For exercise 2, how many buses are in the first group?* (8) *The second group?* (6) *Which group has more?* (the one on the left) *How many buses are in the group that you circled?* (8)

Power Practice
- Check that students understand the two groups they are comparing in each exercise. In exercises 3 and 4, the two groups are separated by a line. In exercises 5, 6, and 7, the groups are on either side of the page (left and right).
- Ask students to think about when it is easier to compare two groups of objects and when it is more difficult. (It's easier when the two groups have very different quantities.) Encourage them to refer to the exercises as they explain.
- Ask: *Is it easier to compare the groups in exercises 3 and 4, or in exercises 5, 6, and 7? Why?* (In general, it should be easier to compare when the objects in the two groups can be matched.)

WHAT IF THE STUDENT CAN'T

Distinguish Right and Left
- Take a penny hike around the school building or on school grounds: At each intersection of two walkways, flip the coin; turn right if it lands on heads, left if it lands tails.
- Try relay races with a new activity for each leg of the race: dribbling a ball with the right hand, hopping on the left foot, and so on.
- With an even number of students, make a grand chain: Form a circle and have every other student face left and the others face right. Each person takes the right hand of the facing person, moving past him or her, then taking the left hand of the next person, and so on.

Complete the Power Practice
- Discuss each incorrect answer. Have the student model counting the buses in each group, recording numbers, and comparing them. Similarly, have the student model matching the buses in one group with those in the comparison group.

USING THE LESSON

Lesson Goal
- Count objects in two groups separately and together.

What the Student Needs to Know
- Count and write numbers to 5.
- Distinguish two groups of objects.

Getting Started
- Find out what students know about addition. Have them explain what it means to add.
- Try this activity with two-sided counters. Place three counters on the desk, with like colors facing up. Ask: *How many counters do I have?* (3) Flip one of the counters over to show a different color. Ask: *Now how many counters do I have?* (3) Point out that there are two groups. Ask: *How many are in each group?* (2 and 1)

What Can I Do?
Read the question and the response. Then read and discuss the example. Count the planes in each group separately first before counting them together. Use manipulatives to demonstrate the concept of putting two groups together and then counting them as one. Emphasize the summary statement that 3 and 1 are 4 in all. Ask:

- *What does the statement "3 and 1 are 4 in all" mean?* (If you have 3 in one group and 1 in another group and you add them together, it makes 4 in all.)
- *What are the two amounts that we are adding?* (3 and 1)
- *If the first group has 1 and the second group has 3, do we still have the same number in all?* (yes) *How could you check?* (count)

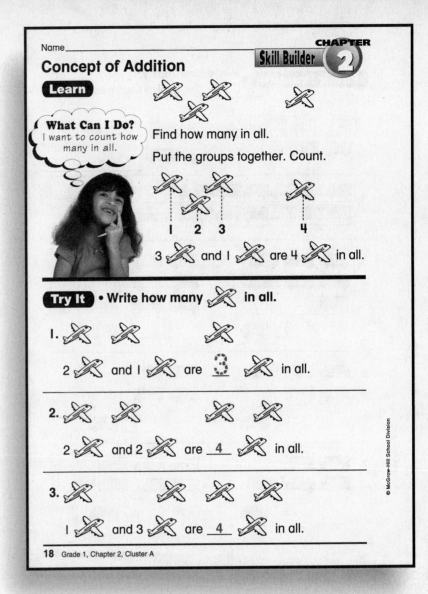

WHAT IF THE STUDENT CAN'T

Count and Write Numbers to 5
- Have student make a 1–5 Counting Book. On each page, the student can draw a picture of a number of objects, such as toys or animals, and then write the number that corresponds to the number of objects in the picture.

Name_____

Power Practice • Write how many ✈ in all.

4. ✈ ✈ _2_ ✈ in all.

5. ✈✈ ✈✈✈ _5_ ✈ in all.

6. ✈ ✈✈ _3_ ✈ in all.

7. ✈✈✈ ✈✈ _5_ ✈ in all.

8. ✈ ✈✈✈✈ _5_ ✈ in all.

9. ✈✈✈ ✈ _4_ ✈ in all.

10. ✈✈✈ ✈✈ _5_ ✈ in all.

11. ✈✈ ✈✈ _4_ ✈ in all.

Grade 1, Chapter 2, Cluster A **19**

USING THE LESSON

Try It
- Read the directions aloud. For exercise 1, have students count the pictures in each group, read the summary statement, and then read the answer. Explain that students should approach each exercise in this way, writing only the total number for both groups together.
- Relate these exercises to everyday situations, for example: *If you borrow 2 books from the library and your friend borrows 2 books, how many can you read together?* (4) *Which exercise is this like?* (exercise 2) *Why?* (It adds two groups of two.)

Power Practice
- As you review the exercises, ask volunteers to provide the summary statements (1 and 1 are 2; 2 and 3 are 5; 1 and 2 are 3; and so on).
- Check that students understand what is meant by addition. Ask: *How do you use counting to help you add?* (You count the items in both groups.)
- Ask students to brainstorm everyday situations when adding is necessary.

Learn with Partners & Parents
- Have students use connecting cubes to form as many different two-color combinations as they can to represent sums of 3, 4, and 5. For example, to show 5, they might attach 2 blue and 3 red, 1 blue and 4 red, and so on. Have them write a summary statement for each combination (2 and 3 is 5, and so on).
- Have students draw pictures or act out scenarios to illustrate the concept of addition.

WHAT IF THE STUDENT CAN'T

Distinguish Two Groups of Objects
- If students tend to count the row of planes in all, without recognizing the two distinct groups that represent addends, try modeling the exercises using two-sided counters or attribute blocks of two distinctly different shapes or sizes.

Complete the Power Practice
- Discuss each incorrect answer. Have the student model counting the planes in each group, saying the summary statement, and writing the total number.

Grade 1, Chapter 2, Cluster A **19**

USING THE LESSON

Lesson Goal
- Identify equivalent amounts.

What the Student Needs to Know
- Count and write numbers 1 to 5.
- Compare different arrangements of the same quantity.
- Recognize the meaning of an equal sign.

Getting Started

Create an array with 4 counters. Using flash cards, place an equal sign, followed by the numeral 4, to the right of the array. Compare the equal sign to a scale: the amount on one side must be the same as the amount on the other side. Add a counter to the pattern and ask:

- *Is this sentence true now?* (no) *Why not?* (Five is more than 4, so the amounts are not equal.)
- *How could I make the sentence true again?* (Change the number to 5 or remove one of the counters.)

Arrange the original 4 counters in another pattern (such as a row or column of 4). Ask:

- *Is the sentence true now?* (yes)
- *How can you tell?* (Count.) Explain that changing the arrangement of objects does not change the quantity.

What Can I Do?

Read the question and the response. Then read and discuss the example. Ask:

- *How many groups do you see?* (3)
- *How would you describe each group?* (Possible answer: A row of 3 cubes lying down, a row of 4 cubes lying down, and a group of 4 cubes standing.)
- *Which two groups have the same number of cubes?* (the groups with 4)
- *What does 4 = 4 mean?* (These two amounts are equal, or the same.)

WHAT IF THE STUDENT CAN'T

Count and Write Numbers 1 to 5

- Let students take turns leading the class in "Simon Says," calling for players to jump four times, clap three times, and so on. Encourage students to count out each set.
- Have students work in groups to create counting posters with the theme of wheeled vehicles. They can draw a unicycle to represent 1, a bicycle for 2, a tricycle for 3, a car for 4, and a new invention for 5. Each person draws a different picture and writes the number several times on a single line.

Compare Different Arrangements of the Same Quantity

- Pair students to practice forming patterns with counters. Provide dominoes, dot number cubes, and other number pattern references. Encourage students to find numerous ways to arrange the counters to represent given numbers.
- Provide connecting cubes that can be joined on all faces. Have students arrange the same number of cubes (no more than 8) in various ways, using three dimensions.

Name_____

Power Practice • Write how many.
Circle numbers that are the same.

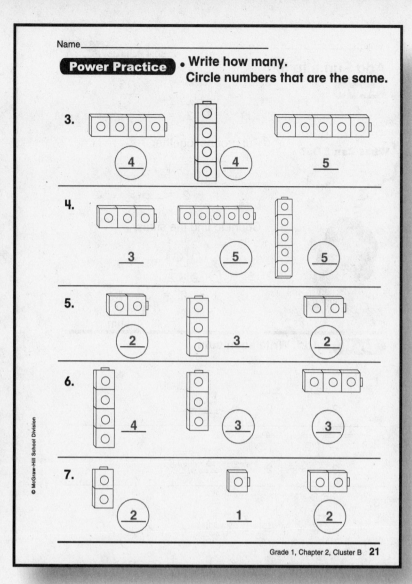

Grade 1, Chapter 2, Cluster B **21**

USING THE LESSON

Try It

- Read the directions aloud and point out the sample answer for exercise 1.
- Check that students understand they are counting the cubes in each group and recording the number on the writing line. Then they can compare the three numbers to see which two are the same.
- Ask volunteers to construct connecting cube models for each exercise. Challenge students to write a number sentence to accompany each exercise. (For exercise 1, it would be 2 = 2; for exercise 2, it would be 3 = 3.)
- Emphasize that the *values* of the two numbers on either side of the equal sign are the same. To highlight this point, erase the number on one side and replace it with an array of dots to show the same number. Allow students to explain in their own words why it does not change the equation.

Power Practice

- Check that students are counting and recording the number of cubes in each exercise and circling the two numbers that are the same.
- Check that students are comparing only the three sets of cubes in the exercise and not the sets of cubes for other exercises.
- You may wish to have students write the equation for each exercise.

WHAT IF THE STUDENT CAN'T

Recognize the Meaning of an Equal Sign

- Explain or review the meaning of an equal sign. Give the student a large equal sign printed on an index card. Ask the student to place from 1–5 counters on one side of the equal sign. Then have the student write the matching number on a square of paper and place it on the other side of the equal sign.

Complete the Power Practice

- Discuss each incorrect answer. Have the student construct models, using connecting cubes, to represent each group in the exercise. He or she can then count each group of cubes, write the number, and circle the two numbers that are the same.

Grade 1, Chapter 2, Cluster B **21**

USING THE LESSON

Lesson Goal
- Add sums to 5.

What the Student Needs to Know
- Recognize the concept of addition.
- Recognize the concept of same.
- Count and write numbers 1 to 5.

Getting Started
- Review the meaning of an equal sign, and introduce the plus sign and the meaning of *sum*. Invite students to explain in their own words what it means to add.
- Survey the class on a topic such as pets. For example, ask who has a cat or a dog at home. Ask children to stand up if they have a cat. As these children remain standing, ask students to stand up if they have a dog. Verbalize the equation: *the number of people who have cats + the number of people who have dogs = the number of people standing.* You may wish to have students count to find the sum, if there aren't too many children standing.

What Can I Do?
- Read the question and the response. Say the equation that is shown in the example: *1 cube plus 2 cubes equals 3 cubes. So we write 3 on the writing line.*
- Explain that the picture of the cube next to the writing line is a reminder that the sum refers to the number of cubes in all. Read the equation again as you point to the pictures: *1 cube plus 2 cubes equals 3 cubes.*
- Ask a volunteer to demonstrate this equation using connecting cubes of two colors.

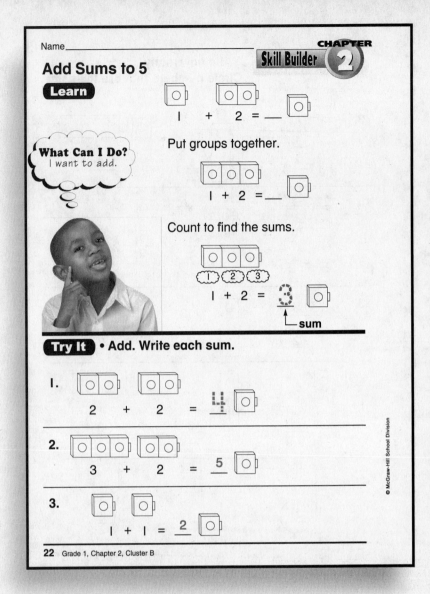

WHAT IF THE STUDENT CAN'T

Recognize the Concept of Addition
- Place two equal trains of connecting cubes on the desk, separated by an equal-sign flash card. Ask the student to split one of the logs in two and separate them with a plus-sign flash card. Read the equation shown. Continue in this way with other sums until the student can read the equations.
- Present everyday examples of addition without using numbers. For example: *A bookcase has two shelves. You want to know how many books you have on both shelves. How might you use addition to find out?*

Recognize the Concept of Same
- Remind the student that the amounts to the left and right of an equal sign must be the same. Ask: *How can you create the same amount without repeating the number?* (Use an arrangement of objects on one side and the number on the other side.) Have the student use from 1–5 counters, an equal sign, and number cards from 1–5 to make up his or her own equation.

Name_____

Power Practice • Add. Write each sum.

4. 2 + 2 = 4

5. 3 + 1 = 4

6. 1 + 4 = 5

7. 2 + 3 = 5

8. 1 + 3 = 4

9. 4 + 1 = 5

10. 2 + 2 = 4

USING THE LESSON

Try It

- Read the directions aloud and remind students to record their answers by writing the sum on the writing line. Check that students understand what is meant by *sum*.
- Provide connecting cubes in colors so that students can model addition.
- Ask volunteers to say the equation for each exercise. (For example, *1 cube plus 1 cube equals 2 cubes.*)
- Ask: *If you have three blue cubes and 2 red cubes, how many do you have in all?* (5) *Which exercise is that like?* (exercise 2)

Power Practice

- Read the directions and remind students to record their answers by writing the sum on the writing line.
- As with the Try It exercises, ask volunteers to say the equations aloud.
- Draw attention to exercise 10. Ask: *In what way is this exercise different from the others?* (The cubes are standing up.) *How is it like exercise 4?* (It shows 2 + 2.)
- Remind students about what they learned in the previous lesson about the equal sign. Ask: *What does the equal sign tell you about the amounts on both sides?* (The amount on one side is the same as the amount on the other side.) *How is 1 + 4 the same as 5?* (When 1 + 4 are put together, there are 5 in all.)
- Challenge students to find other ways to say 5. (4 + 1, 2 + 3, 3 + 2, and so on)

WHAT IF THE STUDENT CAN'T

- Place a train with three connecting cubes beside another train with four connecting cubes. Ask the student to make the trains equal. Repeat with other amounts, focusing on the concept of *same* rather than on sums.

Count and Write Numbers 1 to 5

- Use flash cards and sets of objects to reinforce counting skills.
- Provide picture books for counting, and allow students to read on their own and practice counting.

- Help students draw hopscotch games on the playground using colored chalk, and allow time for them to play.

Complete the Power Practice

- Discuss each incorrect answer. Have the student use connecting cubes to model how to find the sum and then record the answer.

Grade 1, Chapter 2, Cluster B **23**

CHALLENGE

Lesson Goal
- Add sums to 5.

Introducing the Challenge
- Read the directions for the "Addition Town" game and discuss how to play it. First, talk about how to make and use the spinners on page 24. Explain that each player selects two numbers (one from each spinner), finds the sum of those two numbers, and colors any single item on the game board that displays the same number as the sum. Be sure students understand that they can color only one object.
- Select two students to demonstrate the game, and explain the rules as they play.
- Point out that once a building or vehicle has been colored, it can't be colored again.
- Check that students understand how to mark the game board. Ask: *If you spin a 1 and a 2, which number will you look for on the game board?* (3) *How many different 3s are there on the game board?* (7)
- Discuss what should happen when all of the numbers for a particular sum have been colored in. The likely scenario is that the player loses that turn.
- Explain that the player who colors in more buildings and vehicles is the winner.

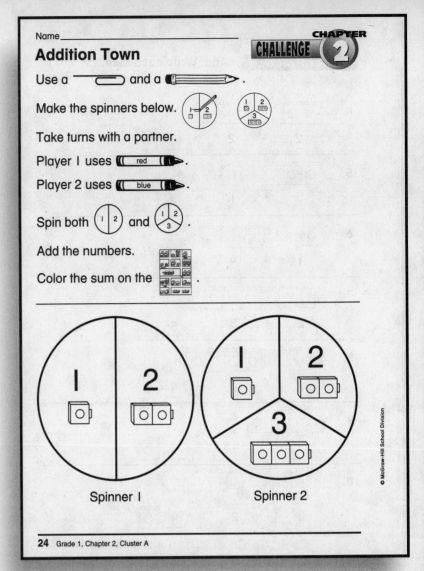

24 Grade 1, Chapter 2, Cluster A

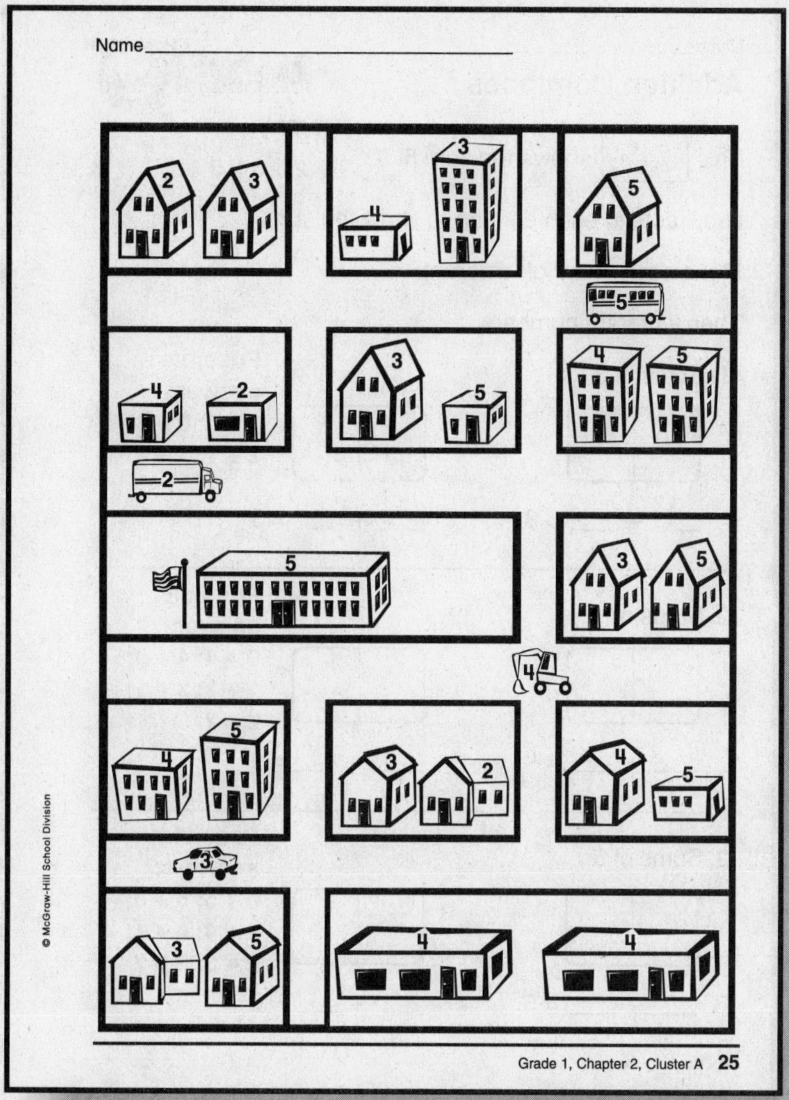

Name_____

Grade 1, Chapter 2, Cluster A **25**

CHALLENGE

- Another option is to declare as the winner the player who colors in the last object. You may wish to have students choose which method to use for declaring the winner.

- Point out that the game board does not have any 1s on it. Ask: *Why isn't it possible to get a sum of 1 in this game?* (You spin once with each spinner.) *What is the least sum possible?* (2) *What is the greatest sum possible?* (5)

Using the Challenge
- Have students work in pairs to play the game.
- Provide paper clips, pencils, scratch paper for sums, and red and blue crayons.
- Check that students understand how to play the game. Check that they understand they do not have to color in the objects on the game board in any particular order.
- If students have difficulty adding, let them use connecting cubes to model the addition.
- Use the game board to review counting. Have students count the the objects labeled with 2, 3, 4, and 5. Make a bar graph to record the data, if you wish.
- Once students are familiar with the game, suggest that they use tally marks to keep track of how many turns it takes them to finish.

Grade 1, Chapter 2, Cluster A **25**

CHALLENGE

Lesson Goal
- Explore different ways to show sums of 3, 4, and 5.

Introducing the Challenge
- Display actual dominoes for the class. Find out what students know about playing dominoes. Explain that there are different games that can be played with dominoes. Allow time for students to explore domino games.
- Review the meaning of *sum*. Explain that this activity uses pictures of dominoes to represent different sums.
- Use connecting cubes in two colors to review different ways to show sums of 3, 4, and 5. For each sum, create a train of cubes in one color to show sums with 0.

Using the Challenge
- Read and discuss the activity directions. Explain that students will record their answers by drawing dots on the blank dominoes and by writing the corresponding numbers on the writing lines. Point out that each exercise focuses on a different sum.
- Review equations (in this case, addition sentences), using the example shown: 2 + 3 = 5. Remind students that the amount to the left of the equal sign must be the same as the amount to the right. That is, 2 + 3 is the same amount as 5.
- Check that students understand how to record their answers.
- After students have completed the activity, have volunteers read the equations aloud. Ask: *How many different ways can you make the sum of 3?* (4) *The sum of 4?* (5) *The sum of 5?* (6) *What pattern do you notice?* (The greater the sum, the more ways there are to form it.)

26 Grade 1, Chapter 2, Cluster B

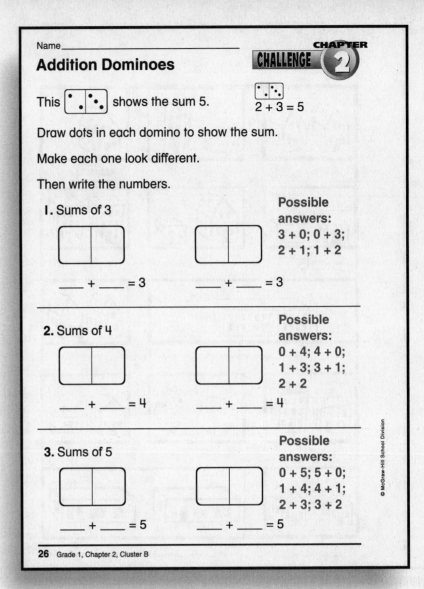

Name_____

More Addition Dominoes

CHALLENGE CHAPTER 2

Draw 2 different dominoes for each sum.

Each space can have only 6 or fewer dots.

6 + 2 = 8   7 + 1 = 8
                ↑
         7 is more than 6.

1. Sums of 6

___ + ___ = 6   ___ + ___ = 6

Possible answers:
0 + 6; 6 + 0;
1 + 5; 5 + 1;
2 + 4; 4 + 2;
3 + 3

2. Sums of 7

___ + ___ = 7   ___ + ___ = 7

Possible answers:
1 + 6; 6 + 1;
2 + 5; 5 + 2;
3 + 4; 4 + 3

3. Sums of 8

___ + ___ = 8   ___ + ___ = 8

Possible answers:
2 + 6; 6 + 2;
3 + 5; 5 + 3;
4 + 4

CHALLENGE

Lesson Goal
- Explore different ways to show sums of 6, 7, and 8.

Introducing the Challenge
- This activity is a continuation of the previous challenge, which explores sums of 3, 4, and 5. Review the meaning of *sum*. Explain to students that this activity will also use pictures of dominoes to stand for different sums.
- Use connecting cubes in two colors to review different ways to show sums of 6, 7, and 8. For each sum, create a train of cubes in one color to show sums with 0.

Using the Challenge
- Read and discuss the activity directions. Focus on the directions that restrict the number of dots to 6 and explain that no more than 6 dots can be used in each space.
- Review equations (in this case, addition sentences), using the example shown: 6 + 2 = 8. Remind students that the amount to the left of the equal sign (in this case, 6 + 2) must be the same as the amount to the right (in this case, 8). Ask: *How can you check to see if this is true?* (count)
- Ask: *What sums of 7 could you not show?* (7 + 0, 0 + 7) *How about 8?* (8 + 0, 0 + 8, 7 + 1, 1 + 7)
- Make a chart of all of the different ways that students showed each sum. Then make a bar graph to record the data, showing how many students selected each way.

Name_____

CHAPTER 3 — What Do I Need To Know?

Numbers to 12

Write each number.

1.  _____

2.  _____

Order Numbers to 12

Write the number that goes in each box.

3.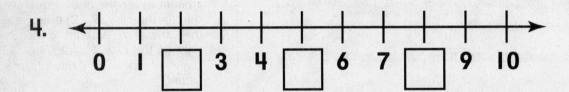

4.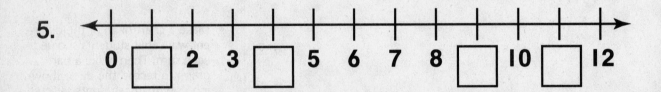

5.

Add Sums to 10

Add. Write each sum.

6.

           6   +   4 = _____

7. 3 + 5 = _____

8.   7
     + 2
     ____

9.   4
     + 1
     ____

Patterns

Write the number that could come next in the pattern.

10. 1 1 2 1 1 2 1 1 2 1 1 ____

CHAPTER 3 PRE-CHAPTER ASSESSMENT

Assessment Goal
This two-page assessment covers skills identified as necessary for success in Chapter 3 Addition Strategies and Facts to 12. The first page assesses the major prerequisite skills for Cluster A. The second page assesses the major prerequisite skills for Cluster B. When the Cluster A and Cluster B prerequisite skills overlap, the skill(s) will be covered in only one section.

Getting Started
- Allow students time to look over the two pages of the assessment. Point out the labels that identify the skills covered.
- Have students find math vocabulary terms used in the assessment. List vocabulary terms on the board as students identify them. If necessary, review the meanings of all essential math vocabulary.

Introducing the Assessment
- Explain to students that these pages will help you know if they are ready to start a new chapter in their math textbooks.
- Students who have transferred from another school may not have been introduced to some of these skills. Encourage students to do their best and assure them you will help them learn any needed skills.

Cluster A Challenge
Those students who demonstrate mastery of the skills on this page will not need to use the reteaching worksheets. Instead, these students can do the Cluster A Challenge found on pages 36-37.

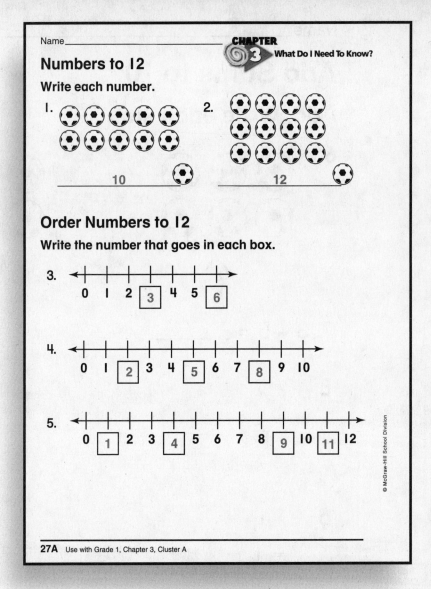

CLUSTER A PREREQUISITE SKILLS

The skills listed in this chart are those identified as major prerequisite skills for students' success in the lessons in Cluster A of the chapter. Each skill is covered by one or more assessment items as shown in the middle column. The right column provides the page numbers for the lessons in this book that reteach the Cluster A prerequisite skills.

Skill Name	Assessment Items	Lesson Pages
Numbers to 12	1-2	28-29
Order Numbers to 12	3-5	30-31

27C Grade 1, Chapter 3, Cluster A

Name_____

Add Sums to 10

Add. Write each sum.

6.

       6   +   4   = __10__

7. 3 + 5 = __8__

8.   7
   + 2
     9

9.   4
   + 1
     5

Patterns

Write the number that could come next in the pattern.

10. 1 1 2 1 1 2 1 1 2 1 1 __2__

© McGraw-Hill School Division

Use with Grade 1, Chapter 3, Cluster B   **27B**

CHAPTER 3 PRE-CHAPTER ASSESSMENT

Alternative Assessment Strategies

- Oral administration of the assessment is appropriate for younger students or those whose native language is not English. Read the skills title and directions one section at a time. Check students' understanding by asking them to tell you how they will do the first exercise in the group.

- For some skill types you may wish to use group administration. In this technique, a small group or pair of students complete the assessment together. Through their discussion, you will be able to decide if supplementary reteaching materials are needed.

Intervention Materials

If students are not successful with the prerequisite skills assessed on these pages, reteaching lessons have been created to help them make the transition into the chapter.

Item correlation charts showing the skills lessons suitable for reteaching the prerequisite skills are found beneath the reproductions of each page of the assessment.

CLUSTER B PREREQUISITE SKILLS

The skills listed in this chart are those identified as major prerequisite skills for students' success in the lessons in Cluster B of the chapter. Each skill is covered by one or more assessment items as shown in the middle column. The right column provides the page numbers for the lessons in this book that reteach the Cluster B prerequisite skills

Skill Name	Assessment Items	Lesson Pages
Add Sums to 10	6-9	32-33
Patterns	10	34-35

Cluster B Challenge

Those students who demonstrate mastery of the skills on this page will not need to use the reteaching worksheets. Instead, these students can do the Cluster B Challenge found on pages 38-39.

Grade 1, Chapter 3, Cluster B   **27D**

USING THE LESSON

Lesson Goal
- Count and write whole numbers through 12

What the Student Needs to Know
- Match objects in one-to-one correspondence
- Read, write, and count numbers to 8
- Identify the same number shown by different arrangements of objects.

Getting Started
- Remind students that they already know how to count to 8. Have them count with you. Say:
- *Let's count to 8, starting with 1. Now let's count: 1-2-3-4-5-6-7-8.*
- *Choose something in your classroom to count that has 8 or fewer items. For example, count the number of windows in the room.* Say: *Let's count the ____ in our room.* Then count them.
- Ask a volunteer to write the number on the chalkboard.
- Repeat the activity with one or two other objects.

What Can I Do?
- Read the question. Then discuss the example.
- Say: *Let's look at these soccer balls. How many do you think there are? Are there more than 8?* Discuss students' responses.
- *Now, let's count these soccer balls. Put your finger on the first ball and count with me.*
- Model placing your finger on each ball as you count.
  1-2-3-4-5-6-7-8. One more than 8 is __?__. (9) One more than 9 is __?__. (10) One more than 10 is __?__. (11) One more than 11 is __?__. (12) How many balls are there? __?__ (12 soccer balls)
- Repeat the activity. This time have students put a finger on each number as you count.
- Say: *Now write how many soccer balls there are.* Have students trace 12.

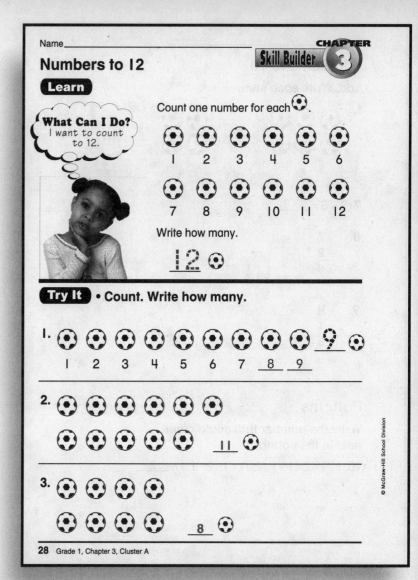

WHAT IF THE STUDENT CAN'T

Match Objects in One-to-One Correspondence
- Start with objects of two-different colors. Have the same number of each (up to 5 of each object). Have the student match the objects into pairs.
- Then have the student match pictured items, such as matching 3 bees to 3 flowers. The student can draw a line to connect the objects.
- Finally have the student match numbers to objects as he or she counts them orally.

Read, Write, and Count Numbers to 8
- Set out 8 objects. Model counting them. Then have the student count. Change the number of objects and have the student count them.
- Have the student match a number card to a given number of objects. Show all 8 cards or just a few cards at a time.
- Then have the student write the number of objects shown. The student can practice forming the numbers by tracing with tracing paper or tracing sample numbers and then writing the numbers independently.

Name_____

Power Practice • Write how many.

4. (7 soccer balls) __7__
5. (10 soccer balls) __10__
6. (12 soccer balls) __12__
7. (9 soccer balls) __9__
8. (8 soccer balls) __8__
9. (11 soccer balls) __11__
10. (12 soccer balls) __12__

© McGraw-Hill School Division

Grade 1, Chapter 3, Cluster A **29**

WHAT IF THE STUDENT CAN'T

Identify the Same Number
- Use counters, connecting cubes, or objects. Name a number from 2–8. Model showing the number in two different ways. For example for 4, you could show 1 row of 4 counters and 2 rows of 2 counters. Then have the student show two different ways to show other numbers.

Complete the Power Practice
- Discuss each incorrect answer. Have the student point to and count the soccer balls and name the number. Then have the student write the correct number.
- If the student has difficulty counting the pictured objects, have the student count actual objects that are set in the same arrangement.

USING THE LESSON

Try It
- Read the directions, Make sure students know what to do.
- Work through the first exercise with the student.
- Ask: *What do we do here?* (Count the soccer balls.)
- *Let's count them together. Put your finger on the first ball. Let's count: 1-2-3-4-5-6. One more than 7 is ___?_.* (8) *So we write 8 on the line. One more than 8 is ___?_.* (9) *So there are __?_ soccer balls in all. What number should we write on the line?* (9)
- Then have students complete exercise 2 and write their answer on the line. Check their work. Repeat for exercise 3.

Power Practice
- Select a few exercises and have volunteers count. Then have the volunteers write the numbers on the chalkboard.
- Direct students' attention to exercise 6. Ask a volunteer to name the number of soccer balls. Then as a group, have students count the soccer balls as they point to them.
- Have students find another exercise that also has 12 soccer balls. (exercise 10) Point out that both exercises have the same number of soccer balls even though the balls are arranged differently.

Learn with Parents & Partners
- Have students make 12 cards with a number from 1–12 on each card. On the back of each card, have students draw dots to match the number on its front.
- Have students use the cards as flash cards. They can name the number or count the dots, depending on the side of the card shown. Encourage them to work as quickly as they can.

Grade 1, Chapter 3, Cluster A **29**

USING THE LESSON

Lesson Goal
- Order whole numbers through 12.

What the Student Needs to Know
- Read, write, and count to 12.
- Identify what comes before and after.

Getting Started
- Draw 12 boxes in a single row on the chalkboard.
- Then count the boxes aloud with students. Remind students to say one number for each object. Say: *Let's count the boxes: 1-2-3-4-5-6-7-8-9-10-11-12. How many boxes are there?* (12)
- Ask volunteers to count the boxes as they point to them.
- Write the number 1 in the first box. Ask students: *What number should I write in the next box?* (2) Have a volunteer write the remaining numbers, 3–12, in the boxes.

What Can I Do?
- Read the question and the response. Then read and discuss the example.
- Point to the top number line. Say: *This number line starts at zero. I can see that some numbers are missing. We can count to find the missing numbers. Let's start at zero.*
- Point to the numbers as you count. Say: *0-1-2-3-____ What comes next?* (4) *Oh yes, 4 comes next. We can write 4 in the box after 3 on the number line. Let's count: 0-1-2-3-4.*
- *Now count with me to find the next missing number. Start at 5: 5-6-____. What comes next?* (7) *We write 7 in the box after 6. Count 0-1-2-3-4-5-6-7.*
- *Count with me to find the last missing number. Start at 8: 8-9-10-11-____. What comes next?* (12) *Write 12 in the box after 11.*
- *Now count with me: 0-1-2-3-4-5-6-7-8-9-10-11-12.*

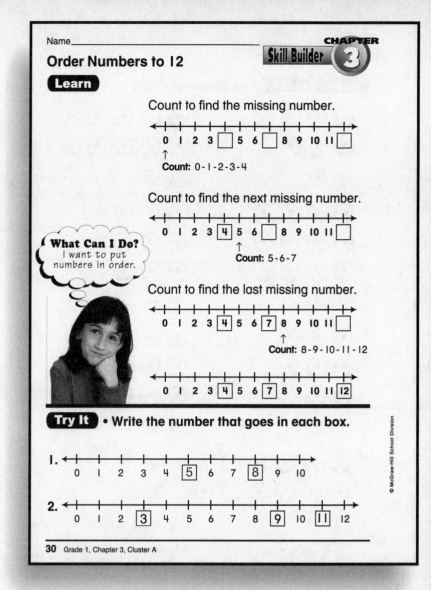

WHAT IF THE STUDENT CAN'T

Read, Write, and Count to 12
- Give the student practice counting real objects or pictured objects. Check that the student is able to count 1–5 objects. Then have the student count 1–8 objects. Finally, ask the student to count 1-12 objects.
- Next, show the student number cards. Have him or her name each number. Again, chunk the cards in these groups: 1–5, 1–8, and 1–12 and review until the student shows little hesitation when shown a card.
- Then have the student match each number card to a group of objects or pictures of objects.
- Have the student practice writing the numbers from 1–12. Start by having the student trace the numbers, then copy them, and then write the numbers without a model. Again work in chunks: 1–5, 6–8, and 9–12.
- Then have the student put all that he or she knows together. Show them a group of objects or pictured objects. Have the student count the objects, name the number, and write the number.

30 Grade 1, Chapter 3, Cluster A

Name_____

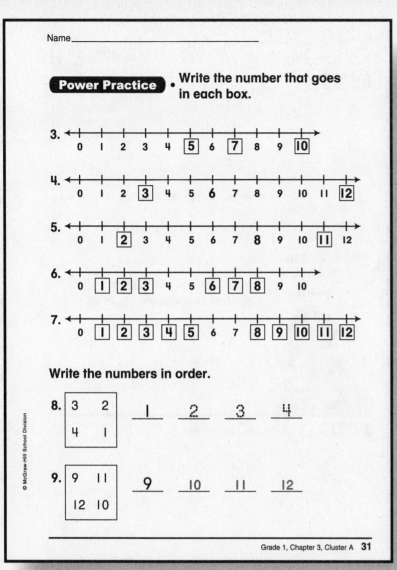

Power Practice • Write the number that goes in each box.

3. 0 1 2 3 4 [5] 6 [7] 8 9 [10]

4. 0 1 2 [3] 4 5 6 7 8 9 10 11 [12]

5. 0 1 [2] 3 4 5 6 7 8 9 10 [11] 12

6. 0 [1] [2] [3] 4 5 [6] [7] [8] 9 10

7. 0 [1] [2] [3] [4] [5] 6 7 [8] [9] [10] [11] [12]

Write the numbers in order.

8. | 3 2 | _1_ _2_ _3_ _4_
   | 4 1 |

9. | 9 11 | _9_ _10_ _11_ _12_
   | 12 10 |

Grade 1, Chapter 3, Cluster A **31**

WHAT IF THE STUDENT CAN'T

Identify What Comes Before and After

- Have a group of students stand in a line facing left. Have each student in turn, except for the first person, point to the person in front of him/her in line and name the person who is before him/her. Each student also points to the person behind and identifies the student who is after him/her.
- Then give each student a number card so that the group of students becomes a human number line. Repeat the same procedure but have each student name the number before and after him/her.

Complete the Power Practice

- Discuss each incorrect answer. For exercises 3–7, have the student point to and count the numbers on the number line. Then have the child write the correct number.
- If any students have difficulty with exercises 8 and 9, have them draw a number line and circle the numbers in the box. Then write the numbers in order.

USING THE LESSON

Try It

- Read the directions and make sure that students know what they are supposed to do.
- Work through the first missing number in excercise 1 with students.
- Ask: *How can we find the first missing number?* (Start counting at 0 and count until you get to the missing number.)
- *Let's count together. Put a finger on each number as you count. Count: 0-1-2-3-4-___. What number comes after 4?* (5) *Write 5 in the first box.*
- Then have children complete exercises 1 and 2 and write their answers in the boxes.
- Ask: *How can you check that your numbers are correct?* (Start at 0 and count.)

Check students' work.

Power Practice

- Read the first set of directions with students and have them complete exercises 3–7.
- Read the second set of directions with children. Say: *The numbers in each box are mixed up. You need to figure out the order of the numbers. Then write the numbers in order.*
- Look at exercise 8. The first number is 1. What number in the box comes next? (2). What number is next? (3) What is the last number? (4) How can you check that the order is correct? (Count the numbers in order.)
- Have students complete exercise 9 in the same way.
- Review selected exercises with students. Discuss how they figured out the number that belonged in each box.

Grade 1, Chapter 3, Cluster A **31**

USING THE LESSON

Lesson Goal
- Add basic facts with sums through 10.

What the Student Needs to Know
- Read, write, and count to 10.
- Identify numbers that are more than or less than another number.
- Understand the concept of addition.

Getting Started
- Put ten unconnected connecting cubes in a container. Ask two students to each reach in and pull out a handful of cubes. Have each student make a cube train and tell how many cubes are in each train.
- Then discuss how many cubes there will be when the cube trains are put together. Record an addition sentence, using an empty box for the sum. Then have a volunteer count the cubes and record the sum in the box.
- Have students read the completed addition sentence.
- Repeat the activity.

What Can I Do?
- Read the question and the responses. Then read and discuss the example.
- Ask: *How many baseballs are in the first group?* (4 baseballs) *How many are in the second group?* (6 baseballs) *What are we trying to find out?* (How many baseballs in all?)
- Ask: *How can you find out how many in all?* (Put the groups together and add.)
- Have students look at the second picture. Ask: *How does this picture show addition?* (The two groups are together.)
- Say: *We know that there are 4 baseballs in this group and we have to add 6 more. How can we find out how many there are in all?* (count on)

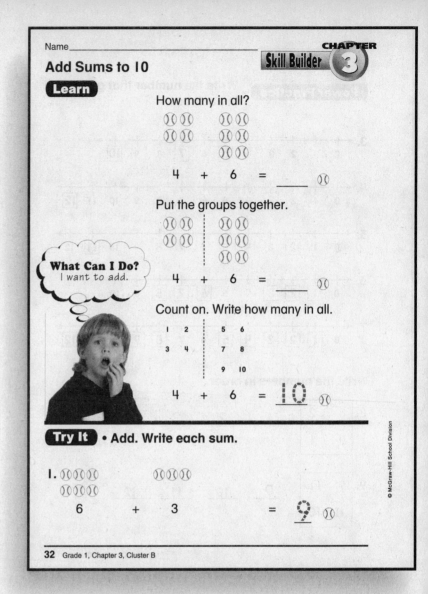

WHAT IF THE STUDENT CAN'T

Read, Write, and Count to 10
- Give the student practice counting real objects and pictured objects. Check that the student is able to count 1–5 objects. Then check that the student can count 1–10 objects.
- Use two of each number flash cards from 1–10 to play games, such as Go Fish and Concentration. Have students name the numbers in the pairs they make.
- Have the student practice copying and writing numbers from 1–10. Then the student can count real or pictured objects, name the number, and write the number.

Identify Numbers That Are More or Less Than Another Number
- Put two colors of connecting cubes in separate containers. Give a handful (5 or less) of each color cube to the student. Have the student count how many are in each group, make cube trains, and tell which group has more and which group has less.
- Repeat the activity several times. Then tell the student two numbers and ask the student which is more and which is less.

Name_____

2. 🥎🥎🥎🥎🥎 🥎🥎🥎🥎🥎
    5 + 5 = 10

3. 4 🥎🥎🥎🥎
   +1 🥎

    5

4. 3 🥎🥎🥎
   +3 🥎🥎🥎

    6

Power Practice • Add. Write each sum.

5. 🥎🥎🥎🥎 🥎
   🥎🥎🥎🥎
   8 + 1 = 9

6. 🥎🥎🥎🥎 🥎🥎🥎🥎
   🥎🥎
   5 + 3 = 8

7. 🥎🥎 🥎🥎🥎🥎
   🥎🥎🥎
   2 + 7 = 9

8. 🥎🥎🥎🥎 🥎
   🥎🥎🥎🥎
   🥎🥎🥎🥎
   9 + 1 = 10

9. 4 🥎🥎🥎🥎
   +4 🥎🥎🥎🥎

    8

10. 5 🥎🥎🥎🥎🥎
    +2 🥎🥎

     7

11. 7 + 3 = 10

12. 4 + 5 = 9

13. 6   14. 3   15. 6   16. 7
    +2   +4   +4   +2
    ---   ---   ---   ---
     8   7   10   9

Grade 1, Chapter 3, Cluster B **33**

USING THE LESSON

- Have students look at the bottom picture. As a group, count the baseballs. Ask: *How many baseballs are there in all?* (10 baseballs) *What is the sum of 4 and 6?* (10)

Try It
- Read the directions and make sure that students know what they are supposed to do.
- Work through the first exercise with children.
- Ask: *How can we find the sum?* (Put the two groups together and count.)
- *Let's count together. Put a finger on each baseball as you count. Count: 1-2-3-4-5-6-7-8-9. So 6 + 3 = ___.* (9) *What number will you write as the sum?* (9)
- Then have students complete exercise 2 and write the sum. Check their work.
- Point out that exercises 3 and 4 are both addition, but they are presented in a column instead of a row. Discuss how students will do the addition. Then have them complete exercises 3 and 4. Check their work.
- If students need support, have connecting cubes or other manipulatives available.

Power Practice
- Read the first set of directions with students and have them complete exercises 5–16.
- Point out that in exercises 11–16 there are no pictures. Discuss strategies students can use to find the sum when there are no pictures. Discuss visualizing the pictures or drawing pictures.

WHAT IF THE STUDENT CAN'T

Understand the Concept of Addition
- Have the student model addition using connecting cubes or two colors. Have the student model one addend with one color and the other addend with the second color. Without recording the addition, ask the student how many cubes there are in all.
- Discuss how the student got the answers. Repeat several times.

Complete the Power Practice
- Discuss each incorrect answer. Have the student use connecting cubes to model any incorrect sums and then correct the answer.

Grade 1, Chapter 3, Cluster B **33**

USING THE LESSON

Lesson Goal
- Determine the next number in a repeating pattern.

What the Student Needs to Know
- Read and write numbers to 9.
- Identify the repeating elements in a pattern.
- Identify the next shape in a pattern.

Getting Started
- Draw this pattern on the chalkboard: square, circle, square, circle, square, circle, ...
- Say: *This pattern repeats. That means that a chunk of the pattern is used over and over. Let's name these shapes and look for the pattern.*
- Point to the shapes as you say them with the students. Allow your voice to adopt the rhythm of the pattern as you say it.
- Ask: *What chunk repeats itself?* (square-circle)
- Ask volunteers to come up and circle each chunk. Then say: *I want to continue this pattern. What shape should I draw next?* (square) *How do you know?* (The last shape was a circle, and the pattern is square-circle.)
- Repeat the activity with a three-shape pattern.

What Can I Do?
- Read the question and the response. Then read and discuss the example.
- Read aloud the pattern with the students as you point to the numbers. Allow the rhythm of the pattern to be evident.
- Ask: *Name the chunk that repeats in this pattern.* (4-5)
- Direct students' attention to the step where the pattern is broken into chunks. Have students identify the chunks.
- Ask: *What number is missing from the last chunk? How can you tell?* (5; the pattern is 4-5 and the last number in the pattern is 4.)

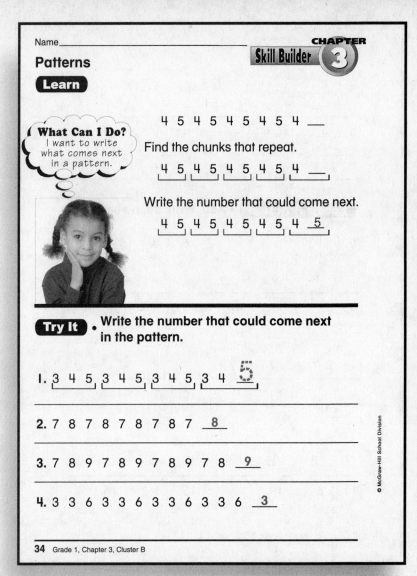

WHAT IF THE STUDENT CAN'T

Read and Write Numbers to 9
- Have the students make a set of flash cards, writing a number from 1–9 on each card.
- Have the student work with a partner who flashes the cards quickly as the student reads the numbers.
- Then have partners take turns dictating numbers from the flash cards while the other partner writes the numbers.
- Once students are comfortable reading and writing numbers, have them make their own number patterns for their partners to complete.

Identify the Repeating Elements in a Pattern.
- Have the student get a red and a blue crayon. Tell the student to draw circles in this pattern: red, blue, red, blue, red, blue,...
- Draw a box around the first chunk of red/blue circles. Explain that this is a chunk. These same two colors repeat over and over.
- Have the student draw boxes around the rest of the chunks.
- Repeat the activity with two different colors. Then try three colors.

Name_____

Power Practice • Write the number that could come next in the pattern.

5. 8 9 8 9 8 9 8 9 8 __9__

6. 2 2 4 2 2 4 2 2 4 2 2 __4__

7. 2 3 5 2 3 5 2 3 5 2 3 __5__

8. 6 7 8 6 7 8 6 7 8 6 7 8 __6__

9. 4 4 8 4 4 8 4 4 8 __4__

10. 5 1 6 5 1 6 5 1 6 5 1 __6__

11. 7 2 9 7 2 9 7 2 9 7 2 __9__

12. 4 5 9 4 5 9 4 5 9 4 5 __9__

13. 3 5 8 3 5 8 3 5 8 __3__

USING THE LESSON

Try It

- Read the directions and make sure that students know what they are supposed to do.
- Work through the first exercise with the students.
- Say: *Let's say this pattern out loud. 3-4-5-3-4-5-3-4-5-3-4-___. What chunk do you think is repeating itself in this pattern?* (3-4-5)
- *Let's look at the pattern to see if all the chunks have 3-4-5. The last chunk is missing a number. Which number is missing?* (5) *How can you tell?* (The chunk is 3-4-5 and the last chunk only has 3-4.)
- *What number should you write on the line?* (5) *Let's read the pattern now and see if 5 works: 3-4-5-3-4-5-3-4-5-3-4-5. Yes, that pattern works.*
- Then have students complete exercises 2–4. Check their work.

Power Practice

- Read the directions with students and havel them complete exercises 5–13.
- Review strategies that students can use to identify the number that could come next in the pattern, such as looking for chunks that repeat, circling the chunks, and saying the pattern aloud and listening for chunks.

WHAT IF THE STUDENT CAN'T

Identify the Next Shape in a Repeating Geometric Pattern.

- Give the student some pattern blocks. Tell the student to make a pattern, such as square, triangle, square, triangle,...
- Have the student identify the chunks in the pattern by separating them so that there is space between each chunk.
- Tell the student to add the next shape in the pattern. Have the student either finish the last chunk or add a new chunk.
- Then have the student say the shapes in the pattern.
- Repeat the activity with other patterns.

Complete the Power Practice

- Discuss each incorrect answer. Have the student identify the chunks in each pattern and draw a box around each repeat of the chunk. Have the student use the complete chunk in the pattern to figure out what the next number could be. Have the student write the correct missing number and then read the pattern aloud.

CHALLENGE

Lesson Goal
- Add using the strategies counting on, adding doubles, adding zero, and adding doubles + 1.

Introducing the Challenge
- Tell students that they are going to play an addition game. Because they are going to spin to find one of the numbers to add, the game will be different every time they play it.
- Explain that once students have added five sums, they will see which player made the greatest sum. That player will be the winner.

Name _____

CHAPTER 3 CHALLENGE

Greatest Sum Game

Play with a partner.

Use a and a ✏️.

Make the spinner below.

To play:

Take turns.

Spin the spinner.

On the game board try to make the greatest sum you can.

At the end of the game, the player with the greatest sum wins.

Spinner sections:
- Same number + 1
- Add 1
- Add 2
- Add 3
- Same number

Game board table:

Player 1 Game 1	Player 2 Game 1
0 + ___ = ___	0 + ___ = ___
1 + ___ = ___	1 + ___ = ___
2 + ___ = ___	2 + ___ = ___
3 + ___ = ___	3 + ___ = ___
4 + ___ = ___	4 + ___ = ___

Player 1 Game 2	Player 2 Game 2
0 + ___ = ___	0 + ___ = ___
1 + ___ = ___	1 + ___ = ___
2 + ___ = ___	2 + ___ = ___
3 + ___ = ___	3 + ___ = ___
4 + ___ = ___	4 + ___ = ___

© McGraw-Hill School Division

Name_____

Greatest Sum Game

Player 1 Game 1	Player 2 Game 1
0 + ____ = ____	0 + ____ = ____
1 + ____ = ____	1 + ____ = ____
2 + ____ = ____	2 + ____ = ____
3 + ____ = ____	3 + ____ = ____
4 + ____ = ____	4 + ____ = ____
Player 1 Game 2	Player 2 Game 2
0 + ____ = ____	0 + ____ = ____
1 + ____ = ____	1 + ____ = ____
2 + ____ = ____	2 + ____ = ____
3 + ____ = ____	3 + ____ = ____
4 + ____ = ____	4 + ____ = ____

CHALLENGE

Using the Challenge

- Give each set of partners a paperclip. Unbend one end of the paperclip so that it makes a pointer. (Make sure students are aware that the end could be sharp.)
- Demonstrate for students how to use the paper clip and a pencil to create a spinner. Put the paper clip in the center of the spinner. Put the pencil in the center of the paper clip and spin the paperclip.
- Have students practice spinning the spinner.
- Give students a game board. Have them fold the paper in half so that they have the Game 1 board in front of them.
- Have students write their names at the top of a column.
- Player 1 spins the spinner. Player 1 adds 1, 2, 3, a double (same number), or a double + 1 (same number + 1) to one of the numbers on the game board. For example, if the player spins "Add 3," the player may choose to add the 3 to the 4 on the game board to get the sum of 7.
- Players take turns until all five additions have been completed.
- Then players circle the greatest sum they made during the game. The player with the greatest sum wins that game.
- Players can then turn the paper over to complete Game 2.
- Note that this is a combination chance/strategy game. Students should eventually recognize that the greatest possible sum will be made by adding the "Same Number + 1" to 4 for a sum of 9. This strategy depends, however, on their spinning spin "Same Number + 1" and on having 4 still open to make the addition.

Grade 1, Chapter 3, Cluster A **37**

CHALLENGE

Lesson Goal
- Match 2 addition sentences that have the same sum.

Introducing the Challenge
- Tell the students that they are going to play an addition game. They are going to play Concentration.
- Explain that in this concentration game, they will turn cards facedown, and take turns flipping over two cards to try to match the sums.
- The winner will be the player who makes the most matches.

Name _____

CHAPTER 3 — CHALLENGE

Addition Concentration

Copy the cards.

Cut them out.

Mix them up.

Turn them face down.

Play with a partner.

To play:

Take turns.

Turn over 2 cards.

Keep cards with the same sum.

Turn over cards that don't have the same sum.

The player with the most cards wins.

3 + 1	2 + 2	2 + 3
4 + 1	3 + 3	4 + 2
5 + 2	6 + 1	4 + 4

38 Grade 1, Chapter 3, Cluster B

Name_____

5 + 3	8 + 1	7 + 2
5 + 5	8 + 2	5 + 1
6 + 0	4 + 3	7 + 0
2 + 6	7 + 1	6 + 3
4 + 5	7 + 3	9 + 1

CHALLENGE

Using the Challenge

- Have the students copy the additions onto index cards. Alternately they can copy the additions onto pieces of paper that have been folded into quarters. Then they can cut out the cards they have made.
- Have students mix up the cards and place them face down on the table.
- The first player turns over two cards. If the cards have the same sum, the player keeps the two cards, and it is the next player's turn.
- If the cards do not have the same sum, then they are turned back over. Encourage students to remember where they saw each sum so that they can turn it over later to make a match when needed.
- When all of the cards have been matched, have students count their cards. The student with more cards wins.

Name_____

CHAPTER 4
What Do I Need To Know?

Numbers to 8

Write how many.

1.

   _____

2.

   _____

More or Fewer

Write the number.
Circle the one that has fewer.

3.

   _____   _____

Add Sums to 8

Add. Write each sum.

4.

      3   +   2   = _____

5.   2

   +4

39A   Use with Grade 1, Chapter 4, Cluster A

Name_____

Same Numbers

**Write each number.
Circle the numbers that are the same.**

6. 

_____ 🦆 _____ 🦆 _____ 🦆

7.

 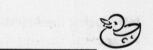

Add Zero

Add. Write each sum.

8.

      4   +   0 = _____ 🦆

9. 8 + 0 = _____   10.   7
                                    + 0

Use with Grade 1, Chapter 4, Cluster B

CHAPTER 4
PRE-CHAPTER ASSESSMENT

Assessment Goal
This two-page assessment covers skills identified as necessary for success in Chapter 4 Subtraction Concepts. The first page assesses the major prerequisite skills for Cluster A. The second page assesses the major prerequisite skills for Cluster B. When the Cluster A and Cluster B prerequisite skills overlap, the skill(s) will be covered in only one section.

Getting Started
- Allow students time to look over the two pages of the assessment. Point out the labels that identify the skills covered.
- Have students find math vocabulary terms used in the assessment. List vocabulary terms on the board as students identify them. If necessary, review the meanings of all essential math vocabulary.

Introducing the Assessment
- Explain to students that these pages will help you know if they are ready to start a new chapter in their math textbooks.
- Students who have transferred from another school may not have been introduced to some of these skills. Encourage students to do their best and assure them you will help them learn any needed skills.

Cluster A Challenge
Those students who demonstrate mastery of the skills on this page will not need to use the reteaching worksheets. Instead, these students can do the Cluster A Challenge found on page 48.

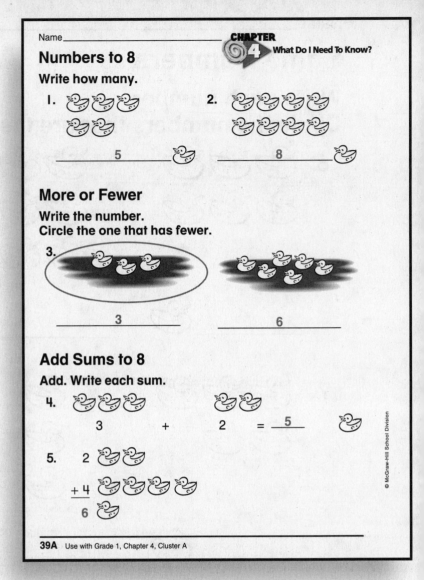

CLUSTER A PREREQUISITE SKILLS

The skills listed in this chart are those identified as major prerequisite skills for students' success in the lessons in Cluster A of the chapter. Each skill is covered by one or more assessment items as shown in the middle column. The right column provides the page numbers for the lessons in this book that reteach the Cluster A prerequisite skills.

Skill Name	Assessment Items	Lesson Pages
Numbers to 8	1–2	40
More or Fewer	3	41
Add Sums to 8	4–5	42–43

39C Grade 1, Chapter 4, Cluster A

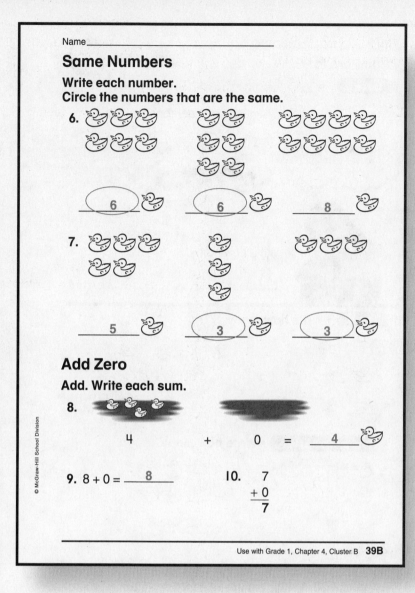

CHAPTER 4 PRE-CHAPTER ASSESSMENT

Alternative Assessment Strategies

- Oral administration of the assessment is appropriate for younger students or those whose native language is not English. Read the skills title and directions one section at a time. Check students' understanding by asking them to tell you how they will do the first exercise in the group.
- For some skill types you may wish to use group administration. In this technique, a small group or pair of students complete the assessment together. Through their discussion, you will be able to decide if supplementary reteaching materials are needed.

Intervention Materials

If students are not successful with the prerequisite skills assessed on these pages, reteaching lessons have been created to help them make the transition into the chapter.

Item correlation charts showing the skills lessons suitable for reteaching the prerequisite skills are found beneath the reproductions of each page of the assessment.

CLUSTER B PREREQUISITE SKILLS

The skills listed in this chart are those identified as major prerequisite skills for students' success in the lessons in Cluster B of the chapter. Each skill is covered by one or more assessment items as shown in the middle column. The right column provides the page numbers for the lessons in this book that reteach the Cluster B prerequisite skills

Skill Name	Assessment Items	Lesson Pages
Same Numbers	6-7	44-45
Add with Zero	8-10	46-47

Cluster B Challenge

Those students who demonstrate mastery of the skills on this page will not need to use the reteaching worksheets. Instead, these students can do the Cluster B Challenge found on page 49.

Grade 1, Chapter 4, Cluster B **39D**

USING THE LESSON

Lesson Goal
- Count and write whole numbers through 8.

What the Student Needs to Know
- Match objects in one-to-one correspondence.
- Read, write, and count numbers to 5.
- Identify the same number shown by different arrangements of objects.

Getting Started
- Ask five students to stand up. Say: *Let's count how many students are standing. 1-2-3-4-5.*
- Have one student sit down. Have volunteers count the children. Repeat until all students are seated.

What Can I Do?
- Read the question and the response. Then read and discuss the example.
- Say: *I see a lot of ducks in this pond. Let's count to see how many there are. Start here at the first duck. It has an arrow with the number 1 on it. Let's count together: 1-2-3-4-5-6-7-8.*
- Ask: *How many ducks are there?* (8 ducks) *What number do we write on the line?* (8)

Try It
- Read the directions and make sure students know what to do.
- Work through the first exercise.
- Ask: *What do we do here?* (Count the ducks)
- *Let's count them together: 1-2-3 ? . The next number is ? .* (4) *How many ducks are there?* (4) *What number do we write on the line?* (4)
- Then have students complete exercise 2.

Power Practice
- Have students complete the practice items. Then review each answer.

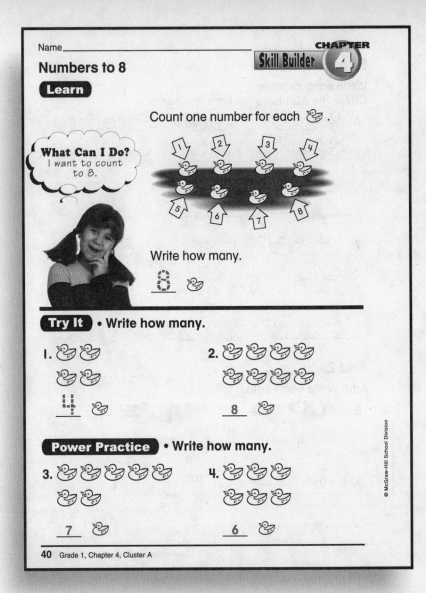

WHAT IF THE STUDENT CAN'T

Match Objects in One-to-One Correspondence
- Give the students some objects, such as buttons or small cars. On index cards, draw button holes or garages. Have the student practice matching one object to its partner.

Read, Write, and Count Numbers to 5
- Have the student practice counting objects and pictures of objects and matching the number counted to the appropriate number card.

Identify the Same Number
- Draw on index cards various arrangements of circles to represent the numbers from 1–8. Make at least two different arrangements for each number.
- Give the student the cards. Tell the student to sort them by matching the pairs that show the same number.

Complete the Power Practice
- Discuss each incorrect answer. Have the student point to and count the ducks and name the number. Then have the student write the correct number.

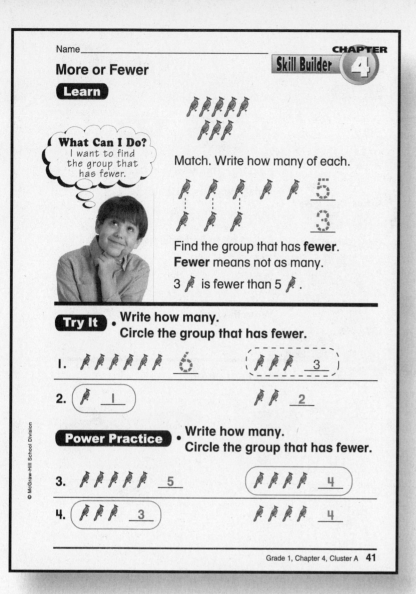

USING THE LESSON

Lesson Goal
- Identify the group that has fewer.

What the Student Needs to Know
- Match objects in one-to-one correspondence.
- Read, write, and count numbers to 5.

Getting Started
- Give each student a pattern block. For each shape give out a different number of blocks.
- Say: *Everyone with circles stand on my left. If you have a square, stand on my right. Are there fewer circles or squares?* Compare the two groups and discuss the answer. Repeat with other combinations.

What Can I Do?
- Read the question and the response. Then read and discuss the example.
- Say: *I see two rows of birds. Which row has fewer birds?*
- First let's count the birds. How many are in the top row? (5) How many in the bottom row? (3) Let's match the birds.
- Model matching the birds. Ask: *Which row has fewer birds?* (the row with 3 birds)

Try It
- Read the directions. Make sure students know what to do.
- Work through the first exercise.
- Ask: *How many birds are in the first group?* (6) *How many are in the second group?* (3) *If you match them, which group has fewer birds?* (3) *Since 3 is fewer than 6, circle 3 birds.*
- Then have students complete exercise 2 and circle the one that is fewer.

Power Practice
- Have students complete the practice items. Then review each answer.

Grade 1, Chapter 4, Cluster A **41**

USING THE LESSON

Lesson Goal
- Add basic facts with sums to 8.

What the Student Needs to Know
- Read, write, and count numbers to 8.
- Identify more or fewer.
- Understand the concept of addition.

Getting Started
- Remind students that they already know how to add sums to 5. Give several students squares of colored paper. Give out 1 red square, 2 blue squares, 3 green squares, and 4 yellow squares.
- Say: *Everyone who has a blue square stand up. How many blue squares are there?* (2) Record the number on the chalkboard.
- Repeat for red squares. Write a plus sign between the 2 numbers and ask: *How many red and blue squares are there?* (3) *How can you check that you are right?* (Put the squares together and count.)
- Repeat for other combinations of squares with sums to 5.

What Can I Do?
- Read the question and the response. Then read and discuss the example.
- Ask: *How many ducks are in the large pond?* (4 ducks) *How many ducks are in the small pond?* (2 ducks) *How can we find out how many there are all together?* (add)
- Remind students that they already know how to count on to add. Say: *Can we count on to find out how many ducks are there in all? Start at 4 and count on two more: 5-6. So 4 ducks and 2 ducks are 6 ducks in all.*
- *What number will you write for the sum?* (6)

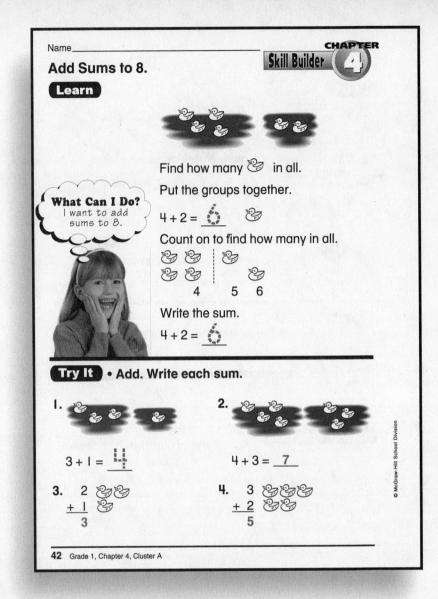

WHAT IF THE STUDENT CAN'T

Read, Write, and Count to 8
- Help the student make a 1–8 number line. Have the student make it bookmark-size to be portable.
- In addition to having the numbers on the line, have the student make arrays of dots or stamps below each number for easy reference.
- Have the student practice reading the numbers and counting the dots/stamps.

Identify More or Fewer
- Have the student make sets of domino cards. Have the student compare the number of dots on the sides of each domino and tell which side has more or fewer.
- Have the student draw one or more sets of "More or Fewer Pictures." Each picture can have two captions, for example: "More cats than dogs" and "Fewer dogs than cats."

Name_____

Power Practice • Add. Write each sum.

5.
   $5 + 1 = \underline{6}$

6.
   $6 + 2 = \underline{8}$

7. 3
   +4
   ─
    7

8. 5
   +3
   ─
    8

9. $6 + 1 = \underline{7}$ 10. $5 + 2 = \underline{7}$

11. $2 + 6 = \underline{8}$ 12. $3 + 1 = \underline{4}$

13. $4 + 4 = \underline{8}$ 14. $7 + 1 = \underline{8}$

15. 1   16. 8   17. 3   18. 2
    +4   +0   +3   +4
    ─   ─   ─   ─
     5   8   6   6

19. 1   20. 2   21. 2   22. 2
    +6   +2   +5   +3
    ─   ─   ─   ─
     7   4   7   5

Grade 1, Chapter 4, Cluster A **43**

WHAT IF THE STUDENT CAN'T

Understand the Concept of Addition

- Write the numbers from 1–4 on separate slips of paper, fold the slips of paper, and place them in a box. Make an identical set of numbered papers and put them in another box.
- Have the student take a paper from one box and model the number with connecting cubes.
- Then have the student take a paper from the other box and model that number as well. Ask the student to say the addition fact as he or she puts the cubes together and finds the sum.
- Repeat the activity several times.

Complete the Power Practice

- Discuss each incorrect answer. Discuss a strategy the student can use to solve each one. Have models available for the student to use.
- Then have the student rework each addition problem and write the correct answer on the line.

USING THE LESSON

Try It

- Read the directions and make sure that students know what they are supposed to do.
- Work through the first exercise with the students. Ask: *How many ducks are in the first pond?* (3 ducks) *How many are in the second pond?* (1 duck) *How can you find the sum of 3 and 1?* (count on 1) *Start at 3 and count 1 more: 3-4. So, what is the sum of 3 + 1?* (4)
- Have students complete exercise 2 and write their answer on the line. Check their work. Point out that exercises 3 and 4 are addition problems that are written in a column instead of a row. Have students complete exercises 3 and 4. Go over students' answers.
- Students who need more support with addition may benefit from using manipulatives.

Power Practice

- Read the directions with the students and make sure they understand what to do.
- Point out that in exercises 9–22 there are no pictures for them to count. They have to remember their addition facts or use a strategy for figuring out the sum.
- Discuss strategies students already know, such as modeling, visualizing, drawing pictures, counting on, and so on.
- Have students complete the exercises. Then select a few exercises and have volunteers show how they found the sum by writing the addition problems on the chalkboard.

Learn with Parents & Partners

- Have students make flash cards for the additions they missed. Have them practice their facts with partners and family members.
- Encourage students to say the entire fact, not just the sum.

Grade 1, Chapter 4, Cluster A **43**

USING THE LESSON

Lesson Goal
- Identify different arrangements of objects as representing the same number.

What the Student Needs to Know
- Read, write, and count numbers to 8.
- Identify same and different shapes.
- Match with one-to-one correspondence.

Getting Started
- Ask four students to bring their chairs to the front of the class. Have them arrange their chairs in a row.
- Say: *Count with me to see how many chairs we have: 1-2-3-4. How many chairs are there?* (4 chairs.)
- Draw 4 chairs or boxes on the chalkboard. Draw them in a row to match the actual chairs. Write the number 4 by the drawing. Say: *This drawing shows the four chairs in a row.*
- Have the four students rearrange their chairs into a 2-by-2 array. Repeat the activity of counting off and representing the chairs on the chalkboard. Repeat with other arrangements.
- Ask: *What do you notice about all of these arrangements?* (The number stays the same but the chairs are grouped differently.)

What Can I Do?
- Read the question and the response. Then read and discuss the example.
- Ask: *How many ducks are in the first pond?* (4 ducks) *How many ducks are in the second pond?* (2 ducks) *How many ducks are in the third pond?* (4 ducks) *Write the number of the ducks on each line.*

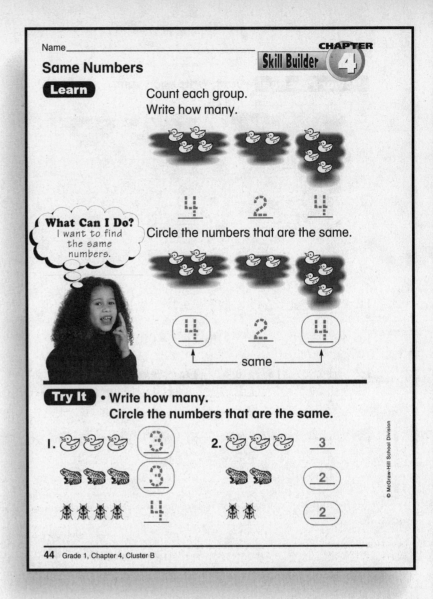

WHAT IF THE STUDENT CAN'T

Read, Write, and Count to 8
- Have the student make a 1–8 counting book. Cut drawing paper in half. Give the student 2 pieces of paper. Have the student fold each paper in half and put them together to make books.
- For each page, show a number flash card for the student to copy on the appropriate page.
- Once all the pages are numbered, have the student draw in a pond the number of animals that matches the number on the page. Check the student's work.
- Encourage the student to decorate the cover of the book and read the book to the class.

Identify Same and Different Shapes
- Give the student some pattern blocks. Ask the student to find two blocks that are the same and to hold them up in the air—one pattern block in each hand.
- Have the student tell how he or she knows the pattern blocks are the same. Repeat with other matches.

Name_____

Power Practice • Write how many. Circle the numbers that are the same.

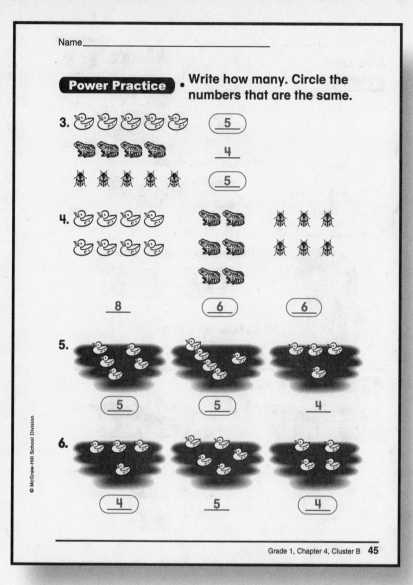

Grade 1, Chapter 4, Cluster B   **45**

WHAT IF THE STUDENT CAN'T

Match with One-to-One Correspondence
- Have the student use 5 counters to cover the 5 ducks in exercise 3. Then have the student move each counter down one by one to cover the frogs. Discuss whether the ducks and frogs match.
- Repeat by having the student match the counters on the ducks to the lady bugs.
- Repeat the activity with exercises 5 and 6. Check that the student is making one-to-one matches.

Complete the Power Practice
- Discuss each incorrect answer. Discuss a strategy the student can use to solve each one. Have the student use connecting cubes to model the numbers if necessary.
- Then have the student redo each incorrect exercise and circle the same numbers.

USING THE LESSON

- *Now how can we tell which ponds have the same number of ducks? (Find the numbers that match.) Is 4 the same as 2? (no) Is 4 the same as 4? (yes) Which numbers are the same? (4 and 4) Circle the numbers that are the same.*

Try It
- Read the directions and make sure that students know what they are supposed to do.
- Work through the first exercise with the students.
- Ask: *How many ducks are there? (3 ducks) How many frogs are there? (3 frogs) How many lady bugs are there? (4 lady bugs) Write each number on the lines.*
- *Read the numbers you wrote. (3-3-4) Which numbers are the same? (3 and 3) Draw a circle around the numbers that are the same.*
- Then have students complete exercise 2 and circle the numbers that are the same. Check their work.
- Students who need more support with identifying the same numbers may benefit from drawing lines to match the pictures one-to-one.

Power Practice
- Read the directions with students and make sure they understand what to do.
- Have the students complete the exercises and check their work.

Grade 1, Chapter 4, Cluster B   **45**

USING THE LESSON

Lesson Goal
- Add zero to a number.

What the Student Needs to Know
- Read, write, and count 0–8.
- Understand the concept of addition.
- Identify the same numbers.

Getting Started
- Ask a question to which the answer will be zero, such as: *How many children in this class have blue hair?* (0 children)
- Ask: *What does zero mean?* (There are none of that thing.)
- Hold out an empty hand. Ask: *How many pennies are in my hand?* (0 pennies)
- Open your other hand. In that hand have 3 pennies. Ask: *How many pennies do I have in this hand?* (3 pennies) *If I put together what I have in both hands, how many pennies will I have in all?* (3 pennies) *Let's count and see: 1-2- ? .* (3)
- Repeat the activity with various numbers of coins in one hand and no coins in the other.

What Can I Do?
- Read the question and the response. Then read and discuss the example.
- Ask: *How many rabbits are in the first box?* (2 rabbits) *How many rabbits are in the second box?* (0 rabbits) *Write on the lines the number of rabbits in each box.*
- Ask: *How can we find the number of rabbits in all?* (Put the rabbits together and count them.) *Let's count the rabbits together: 1–2. How many rabbits are there in all?* (2 rabbits). *Write the sum on the line.*
- Ask: *What happens when you add zero to a number?* (The other addend and the sum are the same.) *Do you think this will always be true? Why?* (Yes, because you aren't adding anything to the number.)

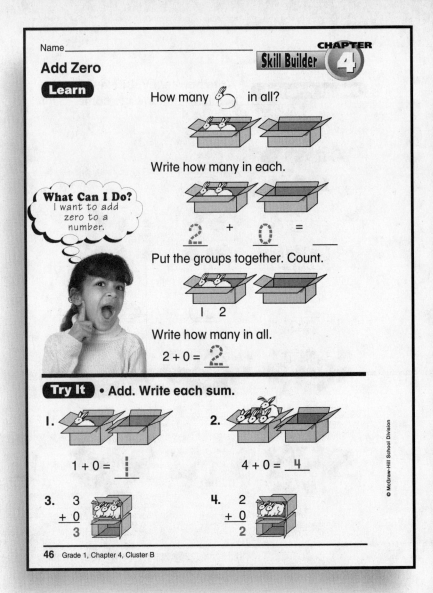

WHAT IF THE STUDENT CAN'T

Read, Write, and Count 0–8
- Make a set of bingo cards. Each card is a 3-by-3-array on which you write in random order the numbers 0–8. Also make 9 flash cards with the numbers from 0–8 on one side and drawings to represent the numbers on the other side. Before the game, use both sides of the cards as flash cards.
- Play the game first by showing the number side of the card. When children win, they must name the winning numbers.
- Show the picture side of the cards for the next game.

Understand the Concept of Addition
- Give the student a piece of paper and some counters. Have the student make addition mats.
- Have the student fold the paper in half, open it, and turn the paper so the fold goes across. Have the student trace the fold with a crayon.
- Next, ask the student to draw a vertical line to divide the top half of the paper into two boxes.
- Now have the student add. For example, the student puts 2 counters and 1 counter in the boxes at the

Name_____

Power Practice • Add. Write each sum.

5.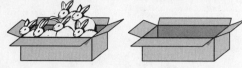
$6 + 0 = \underline{6}$

6.
$8 + 0 = \underline{8}$

7. $\begin{array}{r} 5 \\ +0 \\ \hline 5 \end{array}$ 8. $\begin{array}{r} 7 \\ +0 \\ \hline 7 \end{array}$

9. $2 + 0 = \underline{2}$ 10. $7 + 0 = \underline{7}$

11. $5 + 0 = \underline{5}$ 12. $3 + 0 = \underline{3}$

13. $\begin{array}{r} 4 \\ +0 \\ \hline 4 \end{array}$ 14. $\begin{array}{r} 0 \\ +0 \\ \hline 0 \end{array}$ 15. $\begin{array}{r} 6 \\ +0 \\ \hline 6 \end{array}$ 16. $\begin{array}{r} 1 \\ +0 \\ \hline 1 \end{array}$

USING THE LESSON

Try It

- Read the directions and make sure that students know what they are supposed to do.
- Work through the first exercise with the student. Ask: *How many rabbits are in the first box?* (1 rabbit) *How many rabbits are in the second box?* (0 rabbits) *When we put the rabbits together, how many rabbits are there in all?* (1 rabbit) *Write the sum on the line.*
- Then have students complete exercises 2–4. Point out that in exercises 3 and 4, the addition is shown in a column instead of a row. Check students' work.
- Students who need more support adding with zero may benefit from having manipulatives available.

Power Practice

- Read the directions with students and make sure they understand what to do.
- Point out that in exercises 9–16 there are no pictures. Discuss some strategies students can use to find the sums, such as using models, drawing pictures, and visualizing.
- Have students complete the exercises. Then select a few exercises and have volunteers explain how they completed the problems.

WHAT IF THE STUDENT CAN'T

top. Then the student slides the counters from both boxes into the bottom part of the mat and counts them.

Identify the Same Numbers.

- For each pair of students, make a set of index cards on which you have drawn pairs of different arrangements of circles to represent the numbers 2–8. For each number, there should be 2 index cards.
- Tell each pair to mix up the cards and place them face-down in a pile.
- Have each student take 3 cards. If they have any that show the same number, they

can put them down. When they put down a match, they have to tell what the same number is.

- If they don't have a match, they take turns taking a card from the pile to try to make a match. The winner is the person who has the most matches.

Complete the Power Practice

- Discuss how adding zero means that the sum and the other addend are the same. Have the student use connecting cubes to model the numbers if necessary.

CHALLENGE

Lesson Goal
- Add sums to 8 and complete a picture by connecting the dots in order from the least sum to the greatest sum.

Introducing the Challenge
- Tell students that they are going to solve some additions. Then they will use the sums to complete this dot-to-dot picture.

Using the Challenge
- Tell students to first complete each addition sentence by writing each sum on a line.
- Point out the start of the dot-to-dot. Remind children that they will connect the dots to make a picture. Tell students to look at the sums and connect the dots from the least sum to the greatest sum.
- Once students have completed the dot-to-dot, they may color the picture.

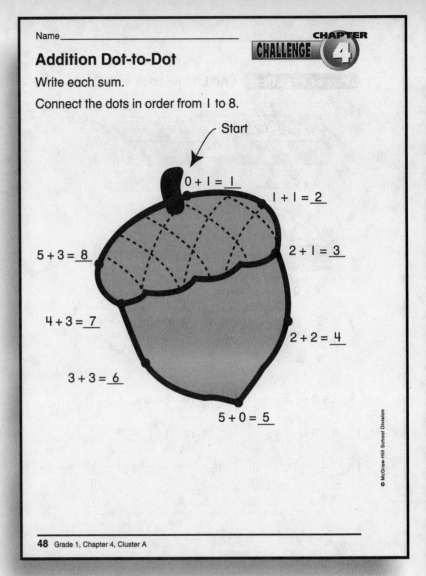

Addition Dot-to-Dot

Write each sum.
Connect the dots in order from 1 to 8.

Start

0 + 1 = __1__

1 + 1 = __2__

5 + 3 = __8__

2 + 1 = __3__

4 + 3 = __7__

2 + 2 = __4__

3 + 3 = __6__

5 + 0 = __5__

48 Grade 1, Chapter 4, Cluster A

Name_____

Number Shapes

CHALLENGE CHAPTER 4

Here are three ways to show 5.

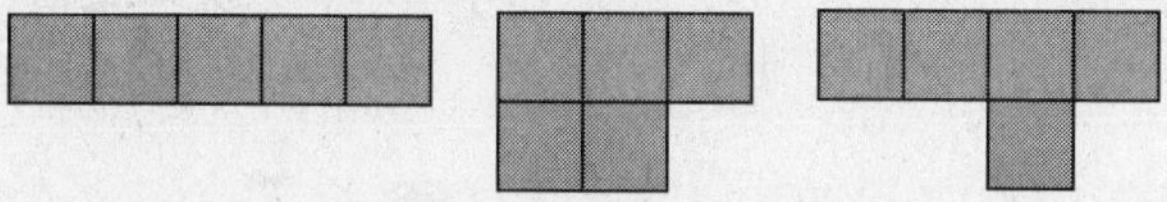

Shade the squares to show each number.

Show three different ways for each number.

1. Show 6.

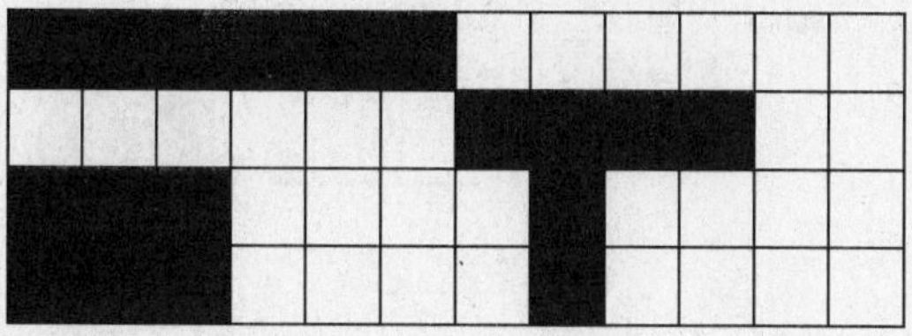

2. Show 8.

Grade 1, Chapter 4, Cluster B **49**

CHALLENGE

Lesson Goal
- Represent the same number using different arrangements of shapes.

Introducing the Challenge
- Remind students that there are different ways to show the same numbers. Tell students that they are going to use squares to show the same number in different ways.

Using the Challenge
- Direct students' attention to the models at the top of the page. Discuss how the models are the same (They all represent the number 5.) and how they are different. (The arrangements of the squares are all different.)
- Explain that students are to shade the boxes on the grids below to show the numbers 6 and 8. Tell students that they need to show each number in three different ways.
- Have students complete exercises 1 and 2.
- If students are eager to try other combinations for these and other numbers, have them work on the back of the paper, or give them grid paper to draw on.

Grade 1, Chapter 4, Cluster B **49**

Name_____

CHAPTER 5 — What Do I Need To Know?

Numbers to 12

Write how many.

1.

_____

Order Numbers to 12

Write the number that goes in each box.

2.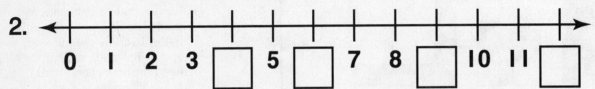

Write the number that comes next when you count.

3. 3, 4, 5, 6, _____

4. 10, 9, 8, 7, _____

Name_____

Add Sums to 12

Add. Write each sum.

5. 9 + 2 = _____

6.   6
   +6
   ___

Subtract from 8

Subtract. Write each difference.

7. 7 − 4 = _____

8.   6
   −5
   ___

9. 8 − 4 = _____

10.   5
    −1
    ___

Use with Grade 1, Chapter 5, Cluster B **49B**

CHAPTER 5 PRE-CHAPTER ASSESSMENT

Assessment Goal

This two-page assessment covers skills identified as necessary for success in Chapter 5 Subtraction Strategies and Facts to 12. The first page assesses the major prerequisite skills for Cluster A. The second page assesses the major prerequisite skills for Cluster B. When the Cluster A and Cluster B prerequisite skills overlap, the skill(s) will be covered in only one section.

Getting Started

- Allow students time to look over the two pages of the assessment. Point out the labels that identify the skills covered.
- Have students find math vocabulary terms used in the assessment. List vocabulary terms on the board as students identify them. If necessary, review the meanings of all essential math vocabulary.

Introducing the Assessment

- Explain to students that these pages will help you know if they are ready to start a new chapter in their math textbooks.
- Students who have transferred from another school may not have been introduced to some of these skills. Encourage students to do their best and assure them you will help them learn any needed skills.

Cluster A Challenge

Those students who demonstrate mastery of the skills on this page will not need to use the reteaching worksheets. Instead, these students can do the Cluster A Challenge found on page 60.

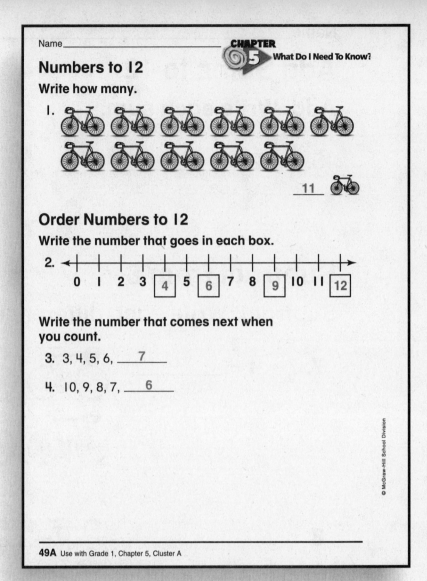

CLUSTER A PREREQUISITE SKILLS

The skills listed in this chart are those identified as major prerequisite skills for students' success in the lessons in Cluster A of the chapter. Each skill is covered by one or more assessment items as shown in the middle column. The right column provides the page numbers for the lessons in this book that reteach the Cluster A prerequisite skills.

Skill Name	Assessment Items	Lesson Pages
Numbers to 12	1	50-51
Order Numbers to 12	2-4	52-53

Name_____

Add Sums to 12
Add. Write each sum.

5. $9 + 2 = \underline{11}$

6. 6
   $\underline{+\,6}$
   12

Subtract from 8
Subtract. Write each difference.

7. $7 - 4 = \underline{3}$

8. 6
   $\underline{-\,5}$
   1

9. $8 - 4 = \underline{4}$

10. 5
    $\underline{-\,1}$
    4

Use with Grade 1, Chapter 5, Cluster B **49B**

CHAPTER 5 PRE-CHAPTER ASSESSMENT

Alternative Assessment Strategies
- Oral administration of the assessment is appropriate for younger students or those whose native language is not English. Read the skills title and directions one section at a time. Check students' understanding by asking them to tell you how they will do the first exercise in the group.
- For some skill types you may wish to use group administration. In this technique, a small group or pair of students complete the assessment together. Through their discussion, you will be able to decide if supplementary reteaching materials are needed.

Intervention Materials
If students are not successful with the prerequisite skills assessed on these pages, reteaching lessons have been created to help them make the transition into the chapter.

Item correlation charts showing the skills lessons suitable for reteaching the prerequisite skills are found beneath the reproductions of each page of the assessment.

Cluster B Challenge
Those students who demonstrate mastery of the skills on this page will not need to use the reteaching worksheets. Instead, these students can do the Cluster B Challenge found on page 61.

CLUSTER B PREREQUISITE SKILLS

The skills listed in this chart are those identified as major prerequisite skills for students' success in the lessons in Cluster B of the chapter. Each skill is covered by one or more assessment items as shown in the middle column. The right column provides the page numbers for the lessons in this book that reteach the Cluster B prerequisite skills

Skill Name	Assessment Items	Lesson Pages
Add Sums to 12	5-6	54-55
Subtract from 8	7-10	56-59

Grade 1, Chapter 5, Cluster B **49D**

USING THE LESSON

Lesson Goal
- Count and write numbers to 12.

What the Student Needs to Know
- Match objects in one-to-one correspondence.
- Read, write, and count numbers to 8.
- Identify the same number shown by different arrangements of objects.

Getting Started
- Make up 12 cards. On each card write a number from 1–12. Give students the cards for numbers 1–8 and review counting.
- Say: *Let's count to 8. Hold up your card when we say your number. Count with me: 1-2-3-4-5-6-7-8.*
- Display the cards for 9–12 on a chalkledge. Point to the appropriate cards as you ask each question. *What comes next after 8? __?__* (9) *After 9? __?__* (10) *After 10? __?__* (11) *After 11? __?__* (12)
- *Let's count to 12. Count with me: 1-2-3-4-5-6-7-8-9-10-11-12.*

What Can I Do?
- Read the question and the response. Then read and discuss the example.
- Say: *We can count the taxis. Remember to start counting on the left at the beginning of each line so you won't lose track of the ones you have already counted.*
- *Put a finger on each one as we count together: 1-2-3-4-5-6-7-8-9-10-11-12. How many taxis are there? __?__* (12) *What number do we write on the line? __?__* (12)

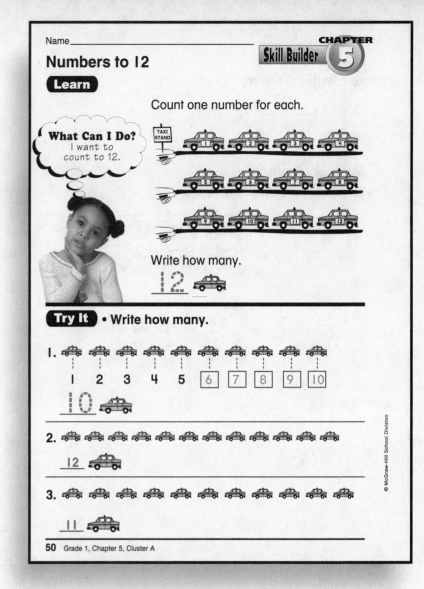

WHAT IF THE STUDENT CAN'T

Match Objects in One-to-One Correspondence
- On sheets of drawing paper, draw a parking lot. Show various sections of the parking lot as having from 2–8 parking spaces per section.
- Ask students to draw one taxi in each parking space.
- When students are done, have them show you the drawing and count the cars in each section.

Read, Write, and Count Numbers to 8
- Set 8 chairs in two rows at the front of the class. Tell students to pretend that these are seats on a bus. Have students count the seats out loud with you.
- Then ask students to act out getting 8 children seated in the bus. Have students count each one.
- Display on a chalkledge flash cards for the numbers from 1–8. Have students read each number as you point to it. Then have students put the appropriate cards on the seats as they count the seats again.
- Have students make their own set of 1–8 flash cards to pratice with at home.

Name_____

Power Practice • Write how many.

4. 🚕🚕🚕🚕🚕🚕🚕
   __7__ 🚕

5. 🚕🚕🚕🚕🚕
   🚕🚕🚕🚕🚕
   __10__ 🚕

6. 🚕🚕🚕🚕
   🚕🚕🚕🚕🚕
   __9__ 🚕

7. 🚕🚕🚕🚕
   🚕🚕🚕🚕
   __11__ 🚕

8. 🚕🚕🚕🚕
   🚕🚕🚕🚕
   🚕🚕🚕🚕
   __12__ 🚕

9. 🚕🚕🚕🚕🚕
   🚕🚕🚕🚕🚕
   __10__ 🚕

10. 🚕🚕🚕🚕🚕🚕
    🚕🚕🚕🚕🚕🚕
    __12__ 🚕

Grade 1, Chapter 5, Cluster A **51**

WHAT IF THE STUDENT CAN'T

Identify the Same Number of Objects
- Have students work with partners.
- Tell each student to make one card for each of the numbers 2–8. Then one student takes his or her number card and draws or stamps a picture to show the number on the other side. The other student then takes his or her card that has the same number, but draws a picture that shows the number in a different way.
- Have students compare their pictures.

Complete the Power Practice
- Discuss each incorrect answer. Discuss a strategy the student can use to solve each one, such as writing the numbers and matching them one-to-one with each object. Have manipulatives available for counting.
- Ask the student to redo each incorrect response and write the correct answer on the line.

USING THE LESSON

Try It
- Read the directions and make sure that students know what they are supposed to do.
- Work through the first exercise with students. Say: *We need to count the taxis. Each taxi needs a number. Start here at 1 and count: 1-2-3-4-5- __?__ What number comes next?* (6) *Write the numbers as you continue counting: 6- __?__* (7-8-9-10)
- Ask: *How many taxis are there?* (10 taxis) *Let's check your work. Go back and put a finger on each taxi as you count it in order: 1-2-3-4-5-6-7-8-9-10. Yes, there are 10 taxis.*
- Have students complete exercises 2 and 3 and review their work.
- Students who need more support with counting may benefit from counting objects.

Power Practice
- Read the directions with students and make sure they understand what to do.
- Have students complete the exercises. Then select a few exercises and have volunteers demonstrate counting the taxis.
- Ask students to look at exercises 5 and 9. Ask them what they notice about the number of taxis. *(They are the same number—10— but they have a different arrangement of the taxis.)*
- Repeat for exercises 8 and 10, which both have 12 taxis.

Grade 1, Chapter 5, Cluster A **51**

USING THE LESSON

Lesson Goal
- Order whole numbers through 12.

What the Student Needs to Know
- Read, write, and count to 12.
- Identify what comes before and after.

Getting Started
- Prepare 13 cards. On each card write a number from 0–12. Display the cards for 0–4 on a chalkledge. Show the cards in a random order.
- Say: *These numbers aren't in the right order. Which number should come first?* (0)
- Ask a volunteer to move the zero card so that it is first. Ask: *Which number goes next?* (1) Continue until all cards are in order. Say: *Let's check that the cards are in order. Count as I point to the cards: 0-1-2-3-4.*
- Repeat the activity. Extend the numbers to 8 and then to 12.

What Can I Do?
- Read the question and the response. Then read and discuss the first example.
- Say: *This number line is missing a number. We can count on to find it. Start at 0 and count with me to find the missing number. 0-1-2-3-4-__?__* (5) *What is the missing number?* (5)
- *Look at the next number line. Let's count together to make sure we are correct: 0-1-2-3-4-5-6-7-8-9-10-11-12. Yes, 5 is the correct number to go in the box.*
- Direct students' attention to the second example. Say: *Since the missing number is near the end, this time we'll count back to find the number. Let's start at 12 and count back together. 12-11-10-__?__* (9) *What is the missing number?* (9) *Let's count back again to make sure that 9 is correct: 12-11-10-9-8-7-6-5-4-3-2-1-0. Yes, 9 is the correct number.*

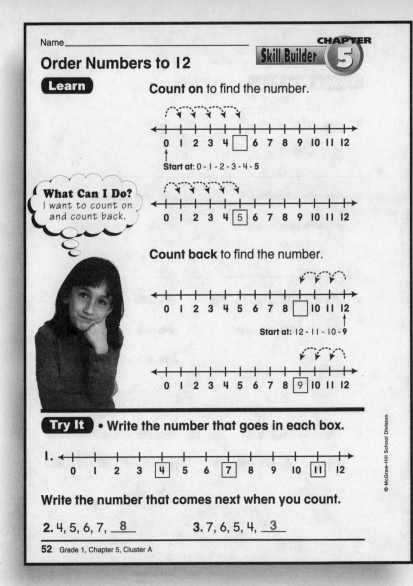

WHAT IF THE STUDENT CAN'T

Read, Write, and Count to 12
- Use flash cards to help students pratice reading numbers to 12. Check that students can easily and quickly read the numbers 1–6 before adding the numbers 7–12 to the set of cards.
- Pull out any cards on which students hesitate when naming the numbers. Have students make their own flash cards for those numbers to use with partners or at home.
- Encourage students to draw a picture or write the beginning sound to help them remember the number the next time they see it on the flash card.
- Show students a number card. Have students count that number of connecting cubes and make a train for the number. Once they have constructed the train, have students take it apart and count the cubes again.
- Make several different cube trains, or have partners make cube trains for each other. Have students count the cubes in the train and write the number on a sticky note and put it on the train.

Name_____

Power Practice • Write the number that goes in each box.

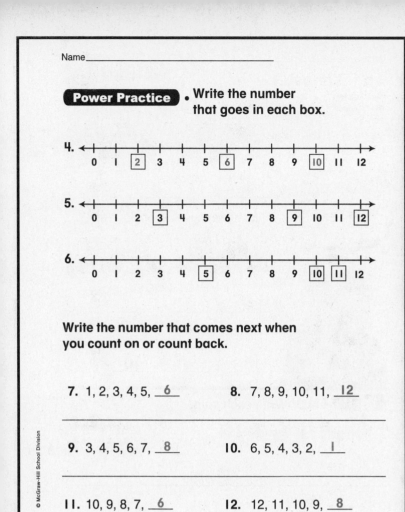

Write the number that comes next when you count on or count back.

7. 1, 2, 3, 4, 5, __6__
8. 7, 8, 9, 10, 11, __12__
9. 3, 4, 5, 6, 7, __8__
10. 6, 5, 4, 3, 2, __1__
11. 10, 9, 8, 7, __6__
12. 12, 11, 10, 9, __8__

Grade 1, Chapter 5, Cluster A **53**

WHAT IF THE STUDENT CAN'T

Identify What Comes Before and After

- Draw a 0–12 number line on the chalkboard. Write only the numbers 0, 5, and 10 on the number line. Draw boxes for the rest of the numbers.
- Ask students to name the numbers that come before and after. For example, start at 0 and ask what comes after it. (1) Have a student add that number to the number line. Then point to 5. Ask what number comes after 5. (6). Then ask for the number that comes before 5. (4) Continue asking questions and having the student fill in the number line until it is complete.

Complete the Power Practice

- Discuss each incorrect answer. Discuss a strategy the students can use to solve each one, such as counting on or counting back. Refer the student to the number lines at the top of page 52 if necessary.
- Then have the student redo each incorrect response and write the correct answer.

USING THE LESSON

Try It

- Read the directions and make sure that students know what they are supposed to do.
- Work through the first exercise with students. Say: *This number line is missing three numbers.* Point to the open box where 4 belongs. Say: *How can you find this missing number?* (count on) *Let's count: 0-1-2-3- __?__ What number comes next?* (4) *That's right. 4 comes next after 3, so write 4 in the box.*
- Now let's find the number that goes in the next box. Let's start counting at 5: *5-6- __?__ What number comes next?* (7)
- This last box is close to the end. We can count back to find it. Start at the end and count back: *12- __?__ . What number comes before 12?* (11) *That's right 11 comes before 12.*
- Let's count the numbers to check our work. Count with me: *0-1-2-3-4-5-6-7-8-9-10-11-12.*
- Read the next set of directions and make sure that students understand what to do. Work through Exercise 2 with students.
- Say: *Let's read the numbers. Tell what comes next: 4-5-6-7- __?__* (8)
- Have students complete Exercise 3. Go over students' work with them.
- Remind students who need more support to use the number lines at the top of the page.

Power Practice

- Read both sets of directions with students and make sure they understand what to do for each practice set.
- Have students complete the exercises. Then select a few exercises and have volunteers demonstrate how they counted on or back to find the missing numbers.

Grade 1, Chapter 5, Cluster A **53**

USING THE LESSON

Lesson Goal
- Add basic facts with sums through 12.

What the Student Needs to Know
- Read, write, and count to 12.
- Identify more than or less than a number.
- Add with sums to 8.

Getting Started
- Have 6 students come up and stand in a line. Have them count off. Write the number of students on the chalkboard.
- Have students stay there as you call up 3 more students. Repeat the activity with the 3 students.
- Say: *Now I want to find out how many students there are here in all. How can I do that?* (add; put the groups together)
- Have the groups move together and count off. Ask a volunteer to write the sum on the chalkboard.

What Can I Do?
- Read the question and the response. Then read and discuss the example.
- Say: *We need to find how many in all. First we need to find how many bikes there are in each bike rack. How many bikes are in the first bike rack?* (5 bikes) *Where will you write that number?* (on the first line before the addition sign)
- Ask: *How many bikes are in the second rack?* (5 bikes) *Where will you write that number?* (on the line just after the addition sign.)
- Ask: *What do you notice about the addition 5 + 5?* (They are doubles.) *How much is 5 + 5?* (10) *How many bikes are there in all?* (10 bikes)

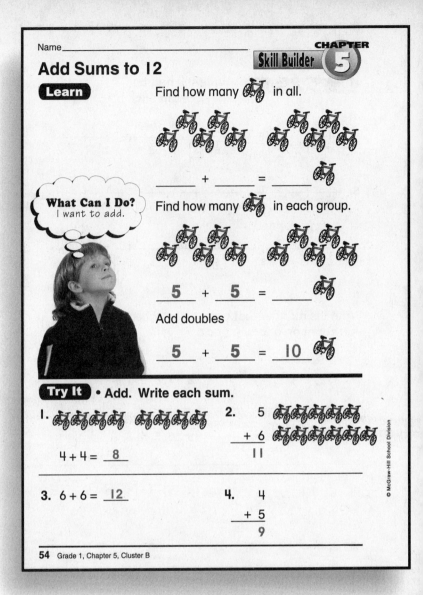

WHAT IF THE STUDENT CAN'T

Read, Write, and Count to 12
- Display a clock face. Count with students from 1–12.
- Have students make a clock faces from a model. Have students make hands for the clocks.
- Ask students to move the big hand to 4 and the little hand to 10. Check their work. Repeat several times with different numbers.
- Dictate a number from 1–12. Have students write it on a sheet of paper. Then have them draw that number of circles. Continue with other numbers.

Identify More Than or Less Than a Number
- Give the student 20 counters. Then draw a 0–12 number line on the chalkboard. Draw a box around 5. Have students model the number. Explain that on the number line, all the numbers to the right of 5 are more than 5. Have them model 6 counters and compare them to the 5 counters. Repeat with other numbers greater than 5.
- Then point out that all the numbers to the left of 5 are less than 5. Have students model 4 and compare it to 5. Repeat for other numbers.

Name_____

Power Practice • Add. Write each sum.

5.   6.

   $9 + 1 = \underline{10}$ $3 + 3 = \underline{6}$

7. 8
   +3
   ‾‾
   11

8. 7
   +2
   ‾‾
    9

9. $6 + 5 = \underline{11}$ 10. $8 + 2 = \underline{10}$

11. $2 + 2 = \underline{4}$ 12. $7 + 5 = \underline{12}$

13. $6 + 1 = \underline{7}$ 14. $6 + 6 = \underline{12}$

15. 9   16. 6   17. 4   18. 8
    +2   +3   +2   +4
    ‾‾   ‾‾   ‾‾   ‾‾
    11   9   6   12

Grade 1, Chapter 5, Cluster B 55

WHAT IF THE STUDENT CAN'T

Add with Sums to 8

- Give students a sheet of grid paper to make addition grids.
- Have students draw a line across the top and down the left the side. Ask them to write an addition sign in the upper left corner. Then have them number from 1–4 across and down.
- Demonstrate to students how to go across and down to find the addends and point out the place to write the sum. Have students complete their grids. Check that the grids are completed correctly.
- Use flash cards for practice with sums to 8. Have students say each sum and then check it on their grids.

Complete the Power Practice

- Discuss each incorrect answer. Discuss a strategy the student can use to solve each one. Have manipulatives available if necessary.
- Then have the student redo each incorrect response and replace it with the correct answer.

USING THE LESSON

Try It

- Read the directions and make sure that students know what they are supposed to do.
- Work through the first exercise with students. Say: *I see 4 bikes in the first bike rack and 4 bikes in the second bike rack. What numbers do you need to add?* (4 + 4) *What is the sum of 4 + 4?* (8) *How many bikes are there in all?* (8 bikes)
- Point out that in exercise 2, the numbers are being added in a column instead of a row. Have students complete the exercise. Review students' work.
- Point out that in exercises 3 and 4, there are no pictures to count. Remind students that they will have to remember the addition facts or use a strategy, such as visualizing or drawing pictures, to find the sums. Have students complete exercises 3 and 4. Review students' work.
- Students who need more support may benefit from having manipulatives available.

Power Practice

- Read the directions with students and make sure they understand what to do.
- Point out that exercises 9–18 do not have pictures to help them count to find the sum. Discuss strategies they can use, such as modeling, visualizing, drawing pictures, and so on.
- Have students complete the exercises. Then select a few exercises and have volunteers demonstrate how they found each sum.

USING THE LESSON

Lesson Goal
- Subtract from 5.

What the Student Needs to Know
- Read, write, and count to 5.
- Identify more than or less than a number.
- Add with sums to 5.

Getting Started
- Have 5 students bring their chairs to the front of the room. Ask all of the students to stand. Ask: *How many students are standing?* (5 students)
- Have 3 students sit down. Ask: *How many students sat down?* (3 students) *How many students are left standing?* (2 students) *Do we have more students standing now than before or fewer students?* (fewer)
- Repeat with other subtractions from 5.

What Can I Do?
- Read the question and the response. Then read and discuss the example.
- Say: *I see 5 trucks getting dirt. Two have already gotten their dirt and are driving away. We need to find how many are left. What subtraction will we do?* (5 – 2)
- We started with 5. We can cross out the two trucks that are driving away to find out how many are left.
- Model crossing out the trucks. Say: *Now let's count the trucks that are left. Count them together as I point to them.* (1-2-3) *How many trucks are left?* (3 trucks) *So what is 5 – 2?* (3) *That's right; 5 – 2 = 3.*

WHAT IF THE STUDENT CAN'T

Read, Write, and Count to 5
- Display flash cards for 1–5 in order. Have students say each number as you point to it in order. Repeat several times.
- Repeat the activity but use the reverse order, and then finally a random order.
- Give students 5 counters. Have students count them aloud.
- Dictate numbers from 1–5. Have students model each number with counters and then write the number on a piece of paper. Provide models for those who need writing support.

Identify More Than or Less Than a Number
- Make up cards for 1–5. On those cards, also show a column of dots corresponding to the number on the card. On another card write "is less than".
- Give one number card to students. Ask two students to compare their cards and find which has fewer dots.
- Tell those students to take the "is less than" card and stand in an order that makes a sentence, such as "3 is less than 4." Have students read the sentence aloud as a group.
- Repeat the activity.

Name_____

Power Practice • Write how many are left.

3. 4 🚛 − 2 🚛 = __2__ 🚛 left.

4. 5 🚛 − 4 🚛 = __1__ 🚛 left.

5. 4 🚛 − 1 🚛 = __3__ 🚛 left.

6. 3 🚛 − 2 🚛 = __1__ 🚛 left.

7. 5 🚛 − 1 🚛 = __4__ 🚛 left.

Grade 1, Chapter 5, Cluster B **57**

WHAT IF THE STUDENT CAN'T

Add with Sums to 5
- Write on the chalkboard addition sentences with sums to 5. Have students model each addition sentence with connecting cubes.
- Use flash cards to reinforce students' recall of basic facts with sums to five. Encourage them to make their own addition flash cards.

Complete the Power Practice
- Discuss each incorrect answer. Help students to find a strategy to use to solve each one. Have manipulatives available if necessary.
- Then have the students redo each incorrect response and replace it with the correct answer.

USING THE LESSON

Try It
- Read the directions and make sure that students know what they are supposed to do.
- Work through the first exercise with students. Ask: *How many trucks are there to begin with?* (3 trucks) *How many trucks are driving away?* (1 truck) *What are you trying to find out?* (how many are left) *What subtraction do you need to do?* (3 − 1) *How can you find out how many are left?* (Cross out the 1 truck that is driving away.) *How many trucks are left?* (2 trucks) *How did you find out?* (We counted the trucks that were left.) *So what is 3 − 1?* (2) That's right; 3 − 1 = 2.
- Have students complete exercise 2. Review students' work.
- Students who need more support may benefit from having manipulatives available.

Power Practice
- Read the directions with students and make sure they understand what to do.
- Discuss with students how they decided how many trucks to cross out. Suggest that if students feel confident that they know the answer, they may write it, and then cross out the trucks to check their answer.
- Have students complete the exercises. Then select a few exercises and ask volunteers to demonstrate how they found each difference.

Grade 1, Chapter 5, Cluster B **57**

USING THE LESSON

Lesson Goal
- Subtract from 8.

What the Student Needs to Know
- Read, write, and count to 8.
- Identify more than or less than a number.
- Add with sums to 8.

Getting Started
- Write a 0–8 number line on the chalkboard.
- Write 4 – 1 on the chalkboard. Ask: *What does the minus sign tell you to do?* (subtract) *What does it mean when you subtract?* (You take something away, so you have less than before.)
- Point to the number 4 on the number line. Say: *Let's do this subtraction together. We start at 4, and we want to subtract 1. Do we count on or count back when we subtract?* (count back) *So, when you count back 1 from 4, what is the answer?* (3)
- Repeat with other subtractions.

What Can I Do?
- Read the question and the response. Then read and discuss the example.
- Say: *There are seven children on the van. The van stops and 2 children get off the van. What do we want to find out?* (how many children are left on the van) *What subtraction do we write to show this?* (7 – 2).
- *We can count back to subtract. Put your finger on the 7 and count back 2. Let's count together: 7 – 6 – ?* (5) *So, what is the difference of 7 – 2?* (5) *How many children are left?* (5 children) *That's right; 7 – 2 = 5.*

WHAT IF THE STUDENT CAN'T

Read, Write, and Count to 8
- As a group say a counting rhyme, such as "1-2-Buckle my Shoe."
- Give students a piece of drawing paper. Have them draw garages with 8 doors. Ask students to count the doors.
- Give students 8 counters. Have students count them.
- Then tell students to pretend that the counters are cars. Ask students to put 5 cars in the garage. Check their work.
- Then have students write a 5 over the fifth garage door. Repeat until all the numbers 1–8 have been used.

Identify More Than or Less Than a Number
- Have students make 1–8 spinners.
- Tell students to work with partners. Each partner spins the spinner and writes down the number spun.
- Have students model the numbers with the connecting cubes. Then have each student state whether his or her number is more than, less than, or the same as the other student's number.

Name_____

Power Practice • Subtract. Write each difference.

7.

8.

6 − 3 = 3

4 − 1 = 3

9.
  8
− 4
  4

10.
  6
− 1
  5

11. 7 − 3 = 4   12. 6 − 4 = 2

13. 3 − 1 = 2   14. 5 − 2 = 3

15. 7   16. 8   17. 5   18. 7
   − 4   − 6   − 4   − 6
     3   2   1   1

Grade 1, Chapter 5, Cluster B 59

USING THE LESSON

Try It

- Read the directions and make sure that students know what they are supposed to do.
- Work through the first exercise with students. Ask: *What subtraction do we need to solve?* (7 − 1) *What does the 7 tell us?* (how many children are on the van) *What does the 1 tell us?* (how many are getting off) *How can we find the answer to 7 − 1?* (count back) *Now count back: 7-__?* (6) *So, what is the difference of 7 − 1?* (6) *How many children are left on the van?* (6 children) *That's right; 7 − 1 = 6.*
- Have students complete exercises 2–6. Point out that some exercises are in columns, not rows, and some do not have pictures to count.
- Discuss strategies students can use, such as drawing pictures, using a number line, and modeling. Then, review students' work with them.
- Students who need more support may benefit from having manipulatives available.

Power Practice

- Read the directions with students and make sure they understand what to do.
- Have students complete the exercises. Then select a few exercises and have volunteers demonstrate how they found each difference.

Learn with Partners & Parents

- Have students make flash cards for any troublesome subtraction facts. Tell students to practice saying each subtraction fact and its difference.
- Each time they can say the fact quickly, they can put a check mark on the card. Once they have 5 check marks on a card, they can "retire" the card. Tell students the card can come out of retirement if that fact ever gets troublesome again.

WHAT IF THE STUDENT CAN'T

Add with Sums to 8

- Give each student a piece of paper. Have them draw lines on the paper to represent two park benches.
- Give students counters and have them model addition facts with sums to 8. For example, say: *On one park bench, there are 3 children. On the other park bench, there are 5 children. Use the counters to show the children on each bench.*
- Ask: *How many children are there all together?* (8) Write that addition sentence. (3 + 5 = 8)
- Have students reverse the order to find the related fact. Then continue with other facts.

Complete the Power Practice

- Discuss each incorrect answer. Discuss a strategy the student can use to solve each one. Have manipulatives available, if necessary.
- Then have the student redo each incorrect response and replace it with the correct answer.

Grade 1, Chapter 5, Cluster B **59**

CHALLENGE

Lesson Goal
- Read, write, and count numbers through 12.

Introducing the Challenge
- Ask students to name the colors of some cars they might see in a parking lot. Point out the parking lot on the paper. Tell students that they are going to color the cars that have parked in this lot.

Using the Challenge
- Read the directions with students Make sure that students have green, red, blue, yellow, and purple crayons.
- If students are unsure of their color words, suggest that they take the crayon as they read each direction and color the crayon on the paper the appropriate color.
- Tell students they can color the cars in any order, just as they would be in a parking lot, or they can use the same colors in a row.
- Remind students that at the end they need to count the purple cars and write the number on the line.
- For students who need some support, suggest that they write the numbers on the cars with pencil before coloring the cars.

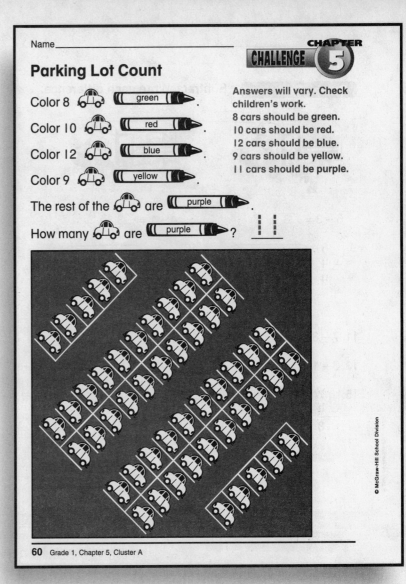

60 Grade 1, Chapter 5, Cluster A

Name_____

Add and Subtract in Circles

Start with the top number.
Look at the sign.
Add or subtract.
Write each missing number.

1.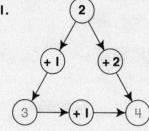

2.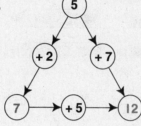

3.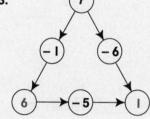

4.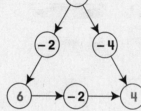

CHALLENGE

Lesson Goal
- Add facts with sums to 8 and subtract from 8.

Introducing the Challenge
- Tell students that they are going to do some number puzzles. Draw a plus sign and a minus sign on the chalkboard. Ask students to tell what each sign tells them to do.

Using the Challenge
- Work through the first example with students. Direct students' attention to the 2 at the top of the triangle. Say: *In these puzzles, you will always start at the top number. Notice that there are two arrows that come down from the top number. Let's follow the arrow on the left.*
- *I see the next circle tells me to add 1. So I need to add the 1 to the 2. How much is 2 + 1? (3) I follow the arrow down. What will I write in the bottom circle? (3)*
- *Now I'll start at the top again and this time I'll follow the arrows down the right side. What does the circle tell me to do? (add 2) What addition do I need to solve? (2 + 2) What is the sum? (4) Where will I write this sum? (in the bottom circle)*
- *Now that I have the two sums in the circles, I can do the addition at the bottom of the triangle. What fact do you see? (3 + 1 = 4)*
- Have students complete exercises 2 – 4. Point out that for exercises 3 and 4, they will have to subtract.

Name_____

Chapter 6 — What Do I Need To Know?

Numbers to 20

Write how many.

1.  _____

2. 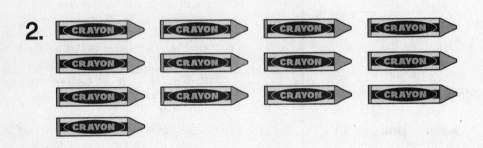 _____

More or Fewer

Circle the one that has more.

3.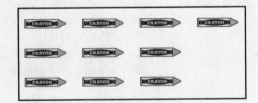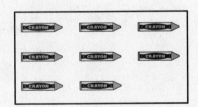

Circle the one that has fewer.

4.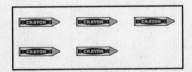

61A Use with Grade 1, Chapter 6, Cluster A

Name_____

Add Sums to 12

Add. Write each sum.

5. 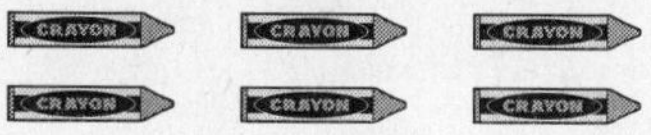 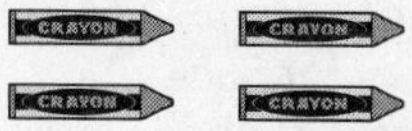   6 + 4 = ____

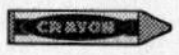

6. 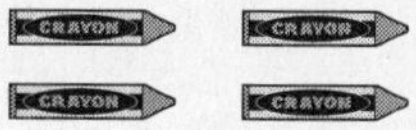 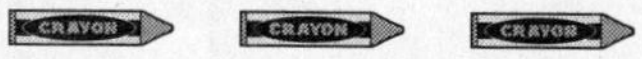   4 + 3 = ____

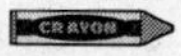

7.   5
    + 6

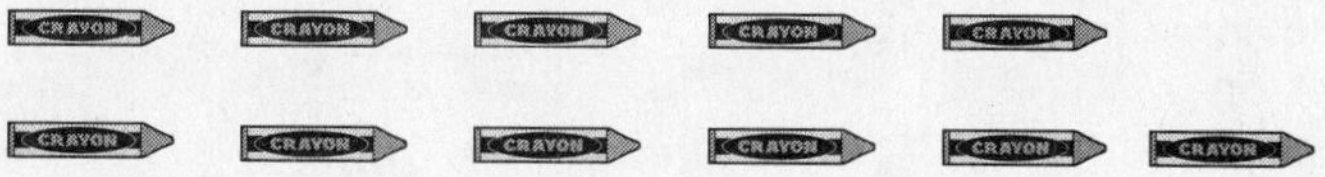

Subtract from 8

Subtract. Write each difference.

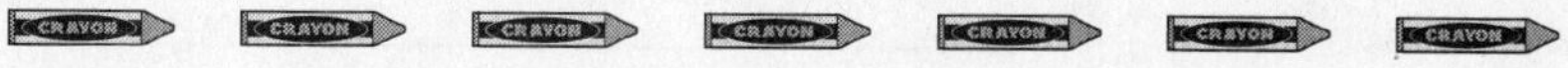

8. 8 − 3 = ____

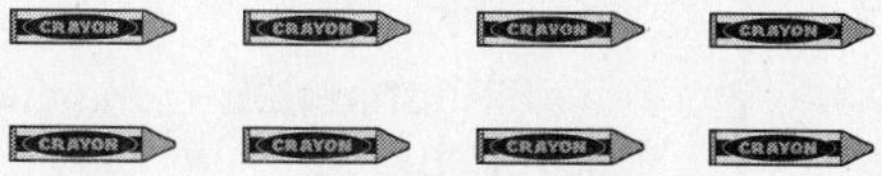

9. 5 − 4 = ____

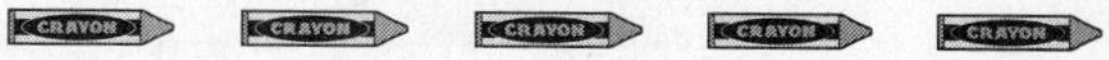

10. 7
    − 2

CHAPTER 6 PRE-CHAPTER ASSESSMENT

Assessment Goal
This two-page assessment covers skills identified as necessary for success in Chapter 6 Data and Graphs. The first page assesses the major prerequisite skills for Cluster A. The second page assesses the major prerequisite skills for Cluster B. When the Cluster A and Cluster B prerequisite skills overlap, the skill(s) will be covered in only one section.

Getting Started
- Allow students time to look over the two pages of the assessment. Point out the labels that identify the skills covered.
- Have students find math vocabulary terms used in the assessment. List vocabulary terms on the board as students identify them. If necessary, review the meanings of all essential math vocabulary.

Introducing the Assessment
- Explain to students that these pages will help you know if they are ready to start a new chapter in their math textbooks.
- Students who have transferred from another school may not have been introduced to some of these skills. Encourage students to do their best and assure them you will help them learn any needed skills.

Cluster A Challenge
Those students who demonstrate mastery of the skills on this page will not need to use the reteaching worksheets. Instead, these students can do the Cluster A Challenge found on page 70.

61C Grade 1, Chapter 6, Cluster A

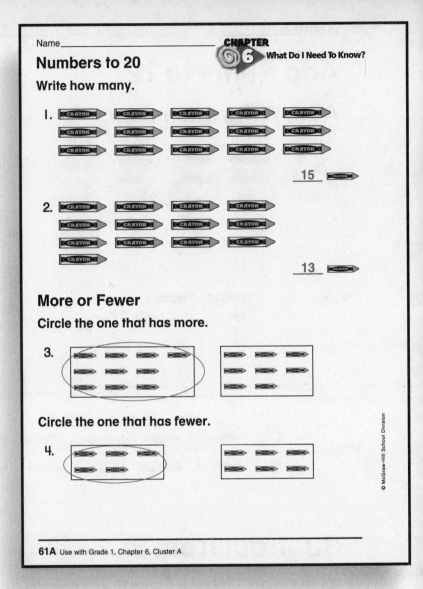

CLUSTER A PREREQUISITE SKILLS

The skills listed in this chart are those identified as major prerequisite skills for students' success in the lessons in Cluster A of the chapter. Each skill is covered by one or more assessment items as shown in the middle column. The right column provides the page numbers for the lessons in this book that reteach the Cluster A prerequisite skills.

Skill Name	Assessment Items	Lesson Pages
Numbers to 20	1-2	62-63
More or Fewer	3-4	64-65

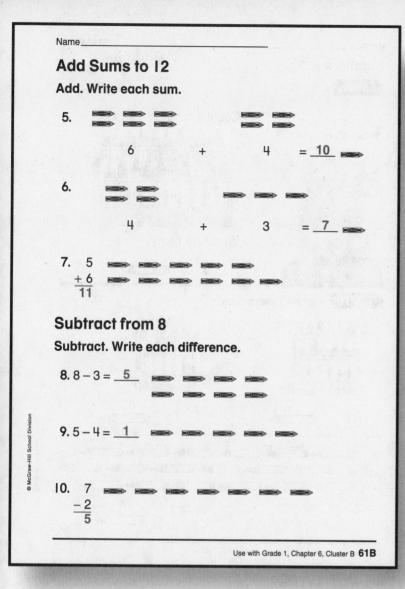

CHAPTER 6 PRE-CHAPTER ASSESSMENT

Alternative Assessment Strategies

- Oral administration of the assessment is appropriate for younger students or those whose native language is not English. Read the skills title and directions one section at a time. Check students' understanding by asking them to tell you how they will do the first exercise in the group.

- For some skill types you may wish to use group administration. In this technique, a small group or pair of students complete the assessment together. Through their discussion, you will be able to decide if supplementary reteaching materials are needed.

Intervention Materials

If students are not successful with the prerequisite skills assessed on these pages, reteaching lessons have been created to help them make the transition into the chapter.

Item correlation charts showing the skills lessons suitable for reteaching the prerequisite skills are found beneath the reproductions of each page of the assessment.

CLUSTER B PREREQUISITE SKILLS

The skills listed in this chart are those identified as major prerequisite skills for students' success in the lessons in Cluster B of the chapter. Each skill is covered by one or more assessment items as shown in the middle column. The right column provides the page numbers for the lessons in this book that reteach the Cluster B prerequisite skills

Skill Name	Assessment Items	Lesson Pages
Add Sums to 12	5-7	66-67
Subtract from 8	8-10	68-69

Cluster B Challenge
Those students who demonstrate mastery of the skills on this page will not need to use the reteaching worksheets. Instead, these students can do the Cluster B Challenge found on page 71.

Grade 1, Chapter 6, Cluster B **61D**

USING THE LESSON

Lesson Goal
- Count and write numbers to 20.

What the Student Needs to Know
- Match objects in one-to-one correspondence.
- Read, write, and count numbers to 12.
- Identify the same number shown by different arrangements of objects.

Getting Started
- Have students stand up. Say: *Let's do some jumping jacks. We can do 12 of them.*
- *Let's take turns saying the numbers as we count the jumping jacks. I'll start with one. For the next one, you'll yell __?__. (two) I'll say three. You'll yell __?__ (four) and so on.*
- *Are you ready? Let's go!* Do the jumping jacks and count off alternately with children.

What Can I Do?
- Read the question and the response. Then read and discuss the example.
- Point to the crayons in the box at the top of the page. Say: *We can count these crayons by going across each row. Remember to count only one number for each crayon.*
- *Put your finger on each crayon as we count. Let's count the top row first. Count out loud with me: 1-2-3-4-5-6-7-8-9-10.*
- Point out the numbers on the crayons. Ask a volunteer to point to each number on the top row of crayons and read them.
- Point to the second row of crayons. Say: *There are more crayons to count. Let's start at the beginning. Count all the crayons this time. Count with me: 1-2-3-4-5-6-7-8-9-10-11-12-13-14-15-16-17-18-19-20.*
- Have a volunteer point to the numbers and read them.
- Ask: *How many crayons are there?* (20) *What number do we write on the line?* (20) *How do we write 20?* (2 and 0)

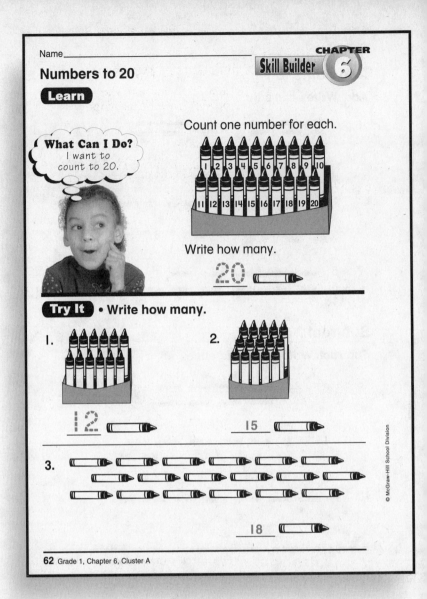

WHAT IF THE STUDENT CAN'T

Match Objects in One-to-One Correspondence
- Give each student a piece of drawing paper and some counters. Have students fold the papers in half.
- Tell students to keep the paper folded and draw one box for each counter they have.
- Demonstrate arranging the counters on the paper and drawing a box for each, or drawing a box and putting a counter in each. Remind students that there needs to be one box for each counter.
- Repeat the activity three more times with different numbers of counters.

Read, Write, and Count Numbers to 12
- Give students 1–12 number lines and 12 counters.
- Demonstrate counting by touching each number as you count aloud slowly from 1–12.
- Then, repeat the activity and have students count the numbers aloud.
- Next, have students take the counters and place one on each number as they say it.
- Then say a number from 1–12. Have students count out the appropriate number of counters and then write the number on the chalkboard.

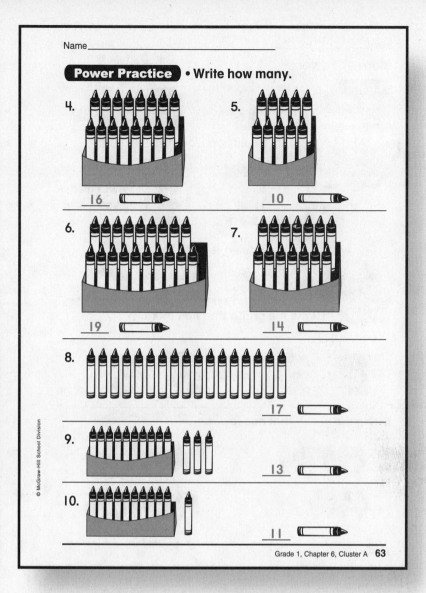

USING THE LESSON

Try It

- Read the directions and make sure that students know what they are supposed to do.
- Work through the first exercise with students.
- Point to the crayons in the first box. Say: *We need to count these crayons. Put your finger on each one and count. Let's start at the back and count one row at a time. Ready? Count: 1-2-3 ?* (4-5-6-7-8-9-10-11-12). *How many crayons are there?* (12 crayons) *What number do you write on the line?* (12)
- Have students complete exercises 2 and 3. Go over students' work with them.
- Encourage students who need more support writing numbers to 20 to refer to the numbers on the crayons at the top of the page.
- Some students who need more support counting to 20 may find it helpful to count objects first.

Power Practice

- Read the directions with students and make sure they understand what to do.
- Have students complete the exercises. Then select a few exercises and ask volunteers to demonstrate how to count the crayons.
- Remind students who need support to refer to the numbered crayons at the top of page 62. Students who need more support counting to 20 may also benefit from having manipulatives to count.

WHAT IF THE STUDENT CAN'T

Identify the Same Number of Objects

- Draw a line down the center of several pieces of paper. Write a different number from 2–12 on each paper. On each paper the same number is written on both sides of the line.
- Give each student 24 counters and one of the papers. Tell students to use the counters to model the number on one side of the paper. Then have them model the same number but in a different way on the other side of the paper.
- Check students' work. Then have students trade papers. Repeat the activity.

Complete the Power Practice

- Discuss each incorrect answer. Ask the student to recount the crayons. Have manipulatives available for counting.
- Then have the student rewrite each incorrect response and write the correct answer on the line.

Grade 1, Chapter 6, Cluster A **63**

USING THE LESSON

Lesson Goal
- Determine which group has more and which has fewer.

What the Student Needs to Know
- Read, write, and count numbers to 8.
- Match objects in one-to-one correspondence.

Getting Started
- Put 5 books in one pile and 4 books in another pile. Use books with different thicknesses so that the taller stack is not necessarily the one with more.
- Ask: *How can we figure out which stack has more books?*
- Discuss students' ideas. Then say: *Let's count the books in each stack and match them. Then we'll see which stack has more.*
- Take the top book from each stack. Say: *One.* Take another book from each stack. Say: *Two.* Continue until there is only 1 book left in the pile that had 5. Ask: *Which stack has more books?* (the stack with 5 books) Say: *So 5 books are more than 4 books.*
- Point to the stack of 4 books. Say: Fewer *means not as many. I can say there are not as many in this stack, or that there are fewer in this stack.*
- Have students name the stack with more or fewer books for other combinations.

What Can I Do?
- Read the question and the response. Then read and discuss the example.
- Point to the stacks of books at the top of the page. Ask: *How can we decide which stack has more books?* (Count the books in each stack and compare the numbers.)

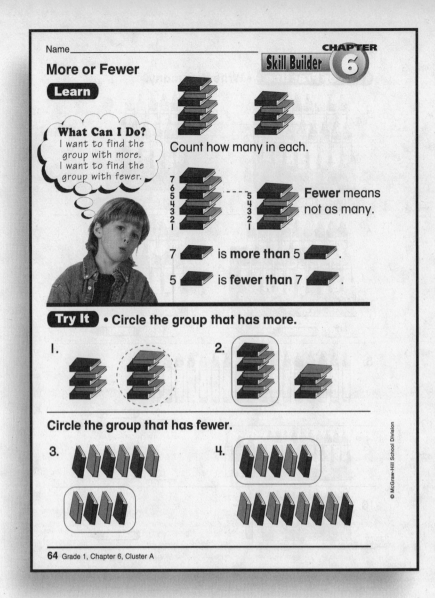

WHAT IF THE STUDENT CAN'T

Read, Write, and Count Numbers to 8
- Write the numbers 1–8 on separate sticky notes. Give a set of sticky notes to each child. Have students arrange the sticky notes in a row on their desks as you say each number. Say: *Find the paper that has 1 written on it. Put it first in a row. Find the paper with the number 2. Put that next in the row.* Continue with the rest of the numbers.
- Once students have all the numbers in a row, have each student read the numbers in order as he or she points to the sticky notes. Then have the students read the numbers and count aloud as a group.
- Give each student between 1 and 8 objects to count. Have students put a sticky note on each object as they count it. Remind students to start at 1 and not to skip any numbers.
- Tell students to write the number of objects they are counting, on a piece of paper. Encourage students to trace the number on the sticky note if they need a model.
- Repeat for other numbers of objects.

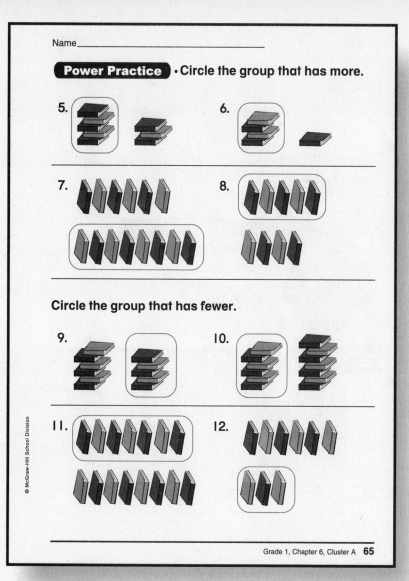

USING THE LESSON

- Say: *Let's count the books in the first stack: 1-2-3-4-5-6-7. How many books are in this stack?* (7) *Now let's count the books in the other stack: 1-2-3-4-5. Did we get to 7?* (no) *So 7 is more than 5.* Ask: *Is 5 fewer than 7?* (yes) *How do you know?* (5 is not as many as 7.)

Try It

- Read the directions for exercises 1 and 2 and make sure that students know what they are supposed to do.
- Work through the first exercise with the students.
- Say: *We need to find out which stack has more. How can we find that out?* (Count the books in each stack and compare the numbers.) *How many books are in the first stack?* (5 books) *How many are in the second stack?* (6 books) *Which is more, 5 or 6?* (6) *Draw a circle around the stack that has more.*
- Have students complete exercise 2. Review students' work with them.
- Read the directions for exercises 3 and 4. Remind students that this time they circle the one that has fewer books—the stack that does not have as many books. Have students complete exercises 3 and 4. Go over students' work with them.
- Encourage students who need more support differentiating between more and fewer to draw lines to match pairs. Then they can see which has more.

Power Practice

- Have students complete the exercises. Then select a few exercises and have volunteers demonstrate counting the books and comparing the stacks.
- Remind students who need support that when they count the books in two stacks, the first stack that has no more books to count is the stack with fewer books.

WHAT IF THE STUDENT CAN'T

Match Objects in One-to-One Correspondence

- Give each student a piece of paper that has been divided into 4 sections. In each section show between 1 and 8 books.
- Tell students to draw 1 bookmark for each book. Then have them draw lines between each book and bookmark.
- Discuss with students how they knew how many bookmarks to draw each time.

Complete the Power Practice

- Discuss each incorrect answer. Discuss a strategy the student can use to solve each one. For example, have connecting cubes available so that the student can use them to model the stacks of books.
- Then have the student redo each incorrect response and circle the correct answer on its line.

USING THE LESSON

Lesson Goal
- Add numbers with sums to 12.

What the Student Needs to Know
- Read, write, and count numbers to 12.
- Identify more than or less than a number.
- Understand the concept of addition.

Getting Started
- Make a set of flash cards for these doubles addition facts: 1+1, 2+2, 3+3, …6+6.
- Arrange the flash cards in order. Say: *Let's practice the doubles facts. They are easy to remember. Let's say them aloud together: 1 and 1 is 2. 2 and 2 is 4.* Continue with the other facts.
- Display the flash cards in order on a chalkledge. Have students say the facts.
- Discuss with students how they remember each of the doubles facts. Then mix up the cards and have volunteers say each fact with its sum.

What Can I Do?
- Read the question and the response. Then read and discuss the example.
- Direct students' attention to the boxes at the top of the page. Ask: *What do we have to do to find out how many pencils?* (add them) *How do we know what to add?* (count the pencils in each box)
- Direct students' attention to the second set of boxes. Say: *Let's count the pencils in the first box. Point to each pencil as we count: 1-2-3-4-5. How many pencils are there?* (5 pencils) *What number do we write on the first line in the addition sentence?* (5)
- Repeat for the other box.
- Say: *Read the addition out loud.* (5 + 5) *What do you notice about it?* (It is a doubles fact.) *How much is 5 + 5?* (10) *How many pencils are there in all?* (10) *What number will you write in the sum?* (10)

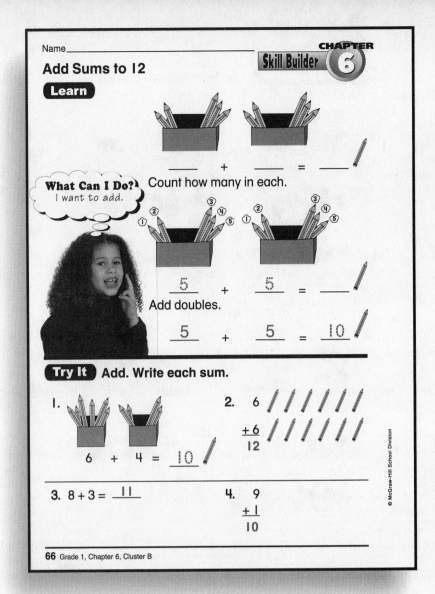

WHAT IF THE STUDENT CAN'T

Read, Write, and Count Numbers to 12
- Number the sections in several egg cartons from 1–12. Give egg cartons and counters to students.
- Have students point to the numbers and read them aloud.
- Tell students to put 10 counters in the sections. Have them start at 1 and put a counter in each section as they count together. Check students' work.
- Then have a volunteer write the number 10 on the chalkboard.
- Repeat the activity with other numbers.

Identify More Than or Less Than a Number
- Make flash cards for the numbers from 1–12. Place them randomly on the chalkledge.
- Point to the 4 card. Say: *I have 4 pieces of chalk. Then draw them on the chalkboard. Say: I want to have more chalk. What number comes after 4?* (5) *Is 5 pieces more than 4 pieces or fewer than 4 pieces?* (more) Draw 5 pieces of chalk. Discuss how students know that 5 is more than 4.

Name_____

Power Practice • Add. Write each sum.

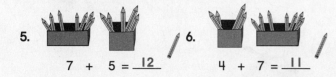

5. 7 + 5 = __12__
6. 4 + 7 = __11__

7. 8
    +2

    10

8. 6
    +3

    9

9. 8 + 4 = __12__
10. 6 + 5 = __11__

11. 7 + 1 = __8__
12. 3 + 3 = __6__

13. 7 + 4 = __11__
14. 9 + 3 = __12__

15. 8   16. 4   17. 5   18. 7
    +1   +4   +4   +3
    ----   ----   ----   ----
     9   8   9   10

Grade 1, Chapter 6, Cluster B **67**

WHAT IF THE STUDENT CAN'T

Understand the Concept of Addition

- Make sentence strips with addition sentences that have sums to 12. Show the addition sentences in word form: "4 and 3 is _?_ "
- Give students counters and a piece of drawing paper. Have them fold the paper in half and open it up. Display the sentence strip and read it aloud.
- Have students model the addition on the mat. Ask them how they can find out how many counters there are in all. (Push the counters together and count them.) Have them find the sum.
- Repeat with other addition sentences.

Complete the Power Practice

- Discuss each incorrect answer. Discuss a strategy the student can use to solve each one, such as counting on, adding doubles, adding doubles + 1, using related facts, modeling, visualizing, and drawing pictures.
- Have connecting cubes available for those students who want to model the addition.
- Have the student redo each incorrect response and replace it with the correct answer.

USING THE LESSON

Try It

- Read the directions and make sure that students know what they are supposed to do.
- Work through the first exercise with students.
- Say: *We need to add the pencils in the boxes. How many pencils are in the first box?* (6 pencils) *How many are in the second box?* (4) *What two numbers do we need to add?* (6 + 4) *Is this a doubles fact?* (no) *How can you find the number of pencils in all?* (Put all the pencils together and count.) *Let's count them together: 1-2-3-4-5-6-7- ? .* (8-9-10) *How many pencils in all?* (10 pencils) *What number do we write in the sum?* (10)
- Point out to students that in exercises 2 and 4 the numbers are added in columns, not rows. Also point out that exercises 3 and 4 do not have pictures. Discuss strategies that students can use for solving the addition problems if they don't recall the facts, for example, drawing pictures, modeling, adding doubles, adding doubles + 1, counting on, and using related facts.
- Have students complete exercises 2–4. Review students' work with them. Students who need more support may benefit from having manipulatives available.

Power Practice

- Read the directions with students. Make sure they understand what to do.
- Have students complete the exercises. Then select a few exercises and have volunteers demonstrate how they completed each addition fact.
- Remind students who need support that they can draw pictures, use or visualize models, or count on.

Grade 1, Chapter 6, Cluster B **67**

USING THE LESSON

Lesson Goal
- Subtract numbers from 8.

What the Student Needs to Know
- Read, write, and count numbers to 8.
- Identify more than or less than a number.
- Add with sums to 8.

Getting Started
- Hold 8 paintbrushes in your hand. Ask a volunteer to take some paintbrushes from you. Have the student count the paintbrushes he or she has.
- Write "8" on the chalkboard. Say: *I started out with 8 paintbrushes. How many paintbrushes did (student's name) take?* (Check students' answer.) *Show the paintbrushes. How many do I have left?* (Check students' answer.)
- Say: *We can write what we just did as a subtraction.* Write the subtraction on the chalkboard. Ask students what the numbers and the symbols in the subtraction stand for.
- Continue the activity. Have one student hold up to 8 paintbrushes and have another student take some. Then have students write the subtraction sentence.

What Can I Do?
- Read the question and the response. Then read and discuss the example.
- Ask: *How many paintbrushes were in the jar to begin with?* (8) *How many paintbrushes is the girl taking from the jar?* (2) *What subtraction will tell us how many paintbrushes are left?* (8 – 2)
- Remind students that different subtraction strategies can be used. Say: *Since we are only subtracting 2, I think it's easiest to count back 2 from 8. Count back 2 with me: 8-7-6. How many paintbrushes are left?* (6 paintbrushes) *What number will we write as the difference?* (6)

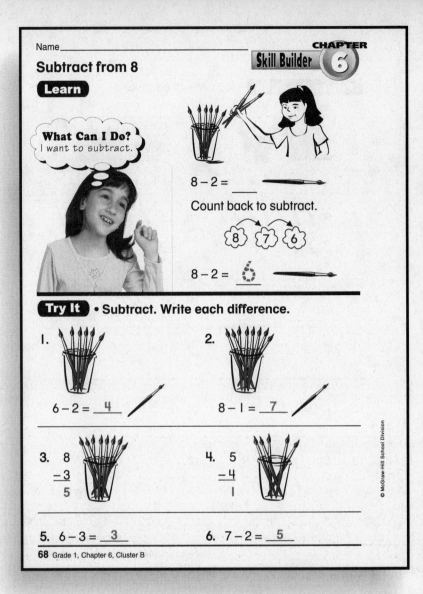

WHAT IF THE STUDENT CAN'T

Read, Write, and Count Numbers to 8
- Use a beanbag. Tell students that you will say "1". You will toss the beanbag to a student who says the next number. That student tosses it to another student who says the next number, and so on to 8. Play the game several times. Then play it by counting back from 8.
- Then toss the beanbag to one student. That student calls out a number from 1–8. The other students write the number and represent it with counters. Repeat the activity several times.

Identify More Than or Less Than a Number
- Write the numbers from 0–8 on slips of paper and place them in a bag. Ask two volunteers to each take a slip of paper from the bag. Have them each read their number aloud.
- Then have each say whether his/her number is more than or less than the other student's number.
- Have students model their numbers with connecting cubes, compare them to check their answers.
- Repeat with other pairs of students.

Name_____

Power Practice • Subtract. Write each difference.

7.
7 − 4 = 3

8.
5 − 3 = 2

9. 6
   −5

    1
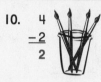

10. 4
    −2

     2

11. 8 − 4 = 4

12. 7 − 1 = 6

13. 6 − 4 = 2

14. 8 − 6 = 2

15. 5 − 2 = 3

16. 4 − 1 = 3

17. 8   18. 6   19. 8   20. 7
    −7   −1   −5   −3
    ---   ---   ---   ---
     1   5   3   4

Grade 1, Chapter 6, Cluster B **69**

WHAT IF THE STUDENT CAN'T

Add with Sums to 8
- Have each student write a number from 1–4 on a slip of paper. Then toss a beanbag to one student who tosses it to another student. Have the two students write on the chalkboard an addition sentence that uses their numbers. Have them add the numbers and write the sum.
- Then have one student toss the beanbag to a third student who checks the answer.
- If the answer is correct, each student gets a point. Count students' points at the end of the game.

Complete the Power Practice
- Discuss each incorrect answer. Discuss a strategy the student can use to solve each one, such as counting back, subtracting doubles, using related facts, modeling, visualizing, and drawing pictures.
- Have connecting cubes available for modeling for those students who need them.
- Have the student redo each incorrect response and replace it with the correct answer.

USING THE LESSON

Try It
- Read the directions and make sure that students know what they are supposed to do.
- Work through the first exercise with students.
- Ask: *How many paintbrushes are in the jar?* (6) *What are you supposed to do?* (subtract 2) *Let's count back from 6 together to find how many are left: 6-5- ? .* (4) *How many are left?* (4) *What number do you write as the difference?* (4)
- Point out to students that some subtractions, like exercises 3 and 4, are written in columns, not rows. Also point out that exercises 5 and 6 do not have pictures.
- Discuss strategies that students can use for solving the subtraction problems, such as crossing out, modeling, and counting back.
- Have students complete exercises 2–6. Review their work with them. Students who need support may benefit from having manipulatives available.

Power Practice
- Read the directions with students. Make sure they understand what to do.
- Have students complete the exercises. Then select a few exercises and have volunteers demonstrate how they completed each subtraction fact.
- Remind students who need support that they can draw pictures, use models, visualize, or count back to solve each subtraction problem.

Learn with Partners & Parents
Have students make tic-tac-toe boards. On each board have them write the numbers 0–8 randomly in the spaces. Have students play tic-tac-toe. To claim a space for their X or O, students must write a subtraction problem whose difference matches the number in the space.

CHALLENGE

Lesson Goal
- Read, write, and count numbers through 20.

Introducing the Challenge
- Ask students if they know what it means when a store takes an inventory. Explain that an inventory is a count of everything that is for sale in a store. Explain that you want students to take an inventory of the people and things in your classroom.

Using the Challenge
- Read the directions with students.
- Tell students that you want them to take a careful count of the number of boys, girls, windows, doors, tables, chairs, bulletin boards, and plants in your classroom right now.
- Explain that students will need an organized way to count since the students may move things around as they count them. Students need to be careful not to count anyone or anything twice. They also need to be careful not to miss anyone or anything.
- Encourage students to check their counts once they have completed them the first time.
- You may also suggest that students compare their counts once they have finished. If there are any differences, they can count the people or things together and agree on a count.

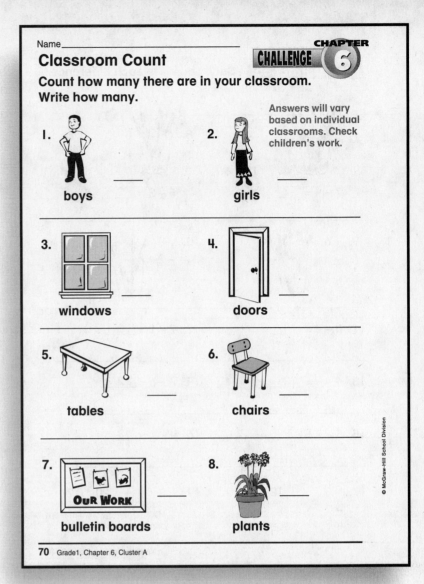

A Difference Maze

Subtract. Write each difference.
Find each difference of 2.
Follow that path through the maze.

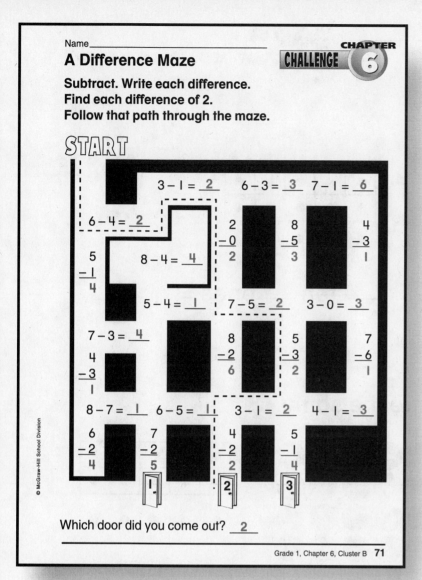

Which door did you come out? __2__

Grade 1, Chapter 6, Cluster B **71**

CHALLENGE

Lesson Goal
- Subtract from 8.

Introducing the Challenge
- Show the maze to students. Ask them if they have ever completed a maze before. Discuss how they figured out how to get through the maze. Explain that this maze is a special maze. They will have to do some subtractions in order to find the correct path out of the maze.

Using the Challenge
- Read the directions with students. Explain that they first need to complete each subtraction and write each difference.

- Tell students that once they have completed all the subtraction problems, they need to find all those subtractions that have a difference of 2.

- Then point out "Start" on the maze. Say: *Put your pencil on "Start". If you go down this path you can either take a turn to 6 – 4 or go straight to 5 – 1. What is the difference of 6 – 4.* (2) *What is the difference of 5 – 1?* (4) *Which path should you take?* (the path with 6 – 4) *Why?* (because it has a difference of 2)

- Have students begin. Remind them to write the number of the flag where they exited the maze.

- For students who get confused with the maze, suggest that they circle all the subtractions that have a difference of 2. Then they can follow the path.

Count to 20

Write how many.

1.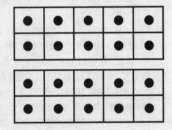

___ ten ___ ones = ___

2.

___ ten ___ ones = ___

Same Number

Write how many.
Circle the numbers that are the same.

3.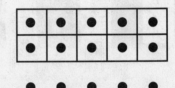

___   ___   ___

More or Fewer

Write how many.
Circle the one that has more.

4.

___   ___

Name_____

Order Numbers to 20

Write each missing number.

5. 0 1 2 3 4 ☐ 6 7 8 9 ☐ 11 12 13 14 ☐ 16 17 18 19 ☐

6. 0 1 ☐ 3 4 5 ☐ 7 ☐ 9 ☐

Patterns

Write what the next number in the pattern could be.

7. 1 2 1 2 1 2 1 2 1 2 ___

8. 1 2 3 4 5 1 2 3 4 5 1 2 3 4 5 1 2 3 4 ___

Compare Numbers to 20

Write > or <.

9. 3 ___ 5

10. 8 ___ 4

CHAPTER 7 PRE-CHAPTER ASSESSMENT

Assessment Goal
This two-page assessment covers skills identified as necessary for success in Chapter 7 Place Value and Patterns. The first page assesses the major prerequisite skills for Cluster A. The second page assesses the major prerequisite skills for Cluster B. When the Cluster A and Cluster B prerequisite skills overlap, the skill(s) will be covered in only one section.

Getting Started
- Allow students time to look over the two pages of the assessment. Point out the labels that identify the skills covered.
- Have students find math vocabulary terms used in the assessment. List vocabulary terms on the board as students identify them. If necessary, review the meanings of all essential math vocabulary.

Introducing the Assessment
- Explain to students that these pages will help you know if they are ready to start a new chapter in their math textbooks.
- Students who have transferred from another school may not have been introduced to some of these skills. Encourage students to do their best and assure them you will help them learn any needed skills.

Cluster A Challenge
Those students who demonstrate mastery of the skills on this page will not need to use the reteaching worksheets. Instead, these students can do the Cluster A Challenge found on page 80.

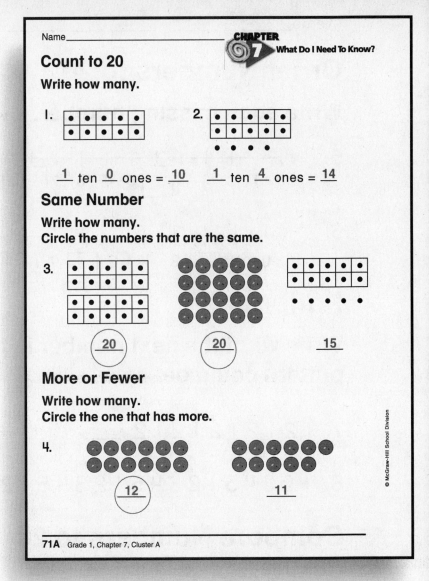

CLUSTER A PREREQUISITE SKILLS

The skills listed in this chart are those identified as major prerequisite skills for students' success in the lessons in Cluster A of the chapter. Each skill is covered by one or more assessment items as shown in the middle column. The right column provides the page numbers for the lessons in this book that reteach the Cluster A prerequisite skills.

Skill Name	Assessment Items	Lesson Pages
Count to 20	1-2	72-73
Same Number	3	74
More or Fewer	4	75

Name_____

Order Numbers to 20

Write each missing number.

5. 0 1 2 3 4 [5] 6 7 8 9 [10] 11 12 13 14 [15] 16 17 18 19 [20]

6. 0 1 [2] 3 4 5 [6] 7 [8] 9 [10]

Patterns

Write what the next number in the pattern could be.

7. 1 2 1 2 1 2 1 2 1 2 __1__

8. 1 2 3 4 5 1 2 3 4 5 1 2 3 4 5 1 2 3 4 __5__

Compare Numbers to 20

Write > or <.

9. 3 __<__ 5

10. 8 __>__ 4

Grade 1, Chapter 7, Cluster B **71B**

CHAPTER 7 PRE-CHAPTER ASSESSMENT

Alternative Assessment Strategies
- Oral administration of the assessment is appropriate for younger students or those whose native language is not English. Read the skills title and directions one section at a time. Check students' understanding by asking them to tell you how they will do the first exercise in the group.
- For some skill types you may wish to use group administration. In this technique, a small group or pair of students complete the assessment together. Through their discussion, you will be able to decide if supplementary reteaching materials are needed.

Intervention Materials
If students are not successful with the prerequisite skills assessed on these pages, reteaching lessons have been created to help them make the transition into the chapter.
Item correlation charts showing the skills lessons suitable for reteaching the prerequisite skills are found beneath the reproductions of each page of the assessment.

CLUSTER B PREREQUISITE SKILLS

The skills listed in this chart are those identified as major prerequisite skills for students' success in the lessons in Cluster B of the chapter. Each skill is covered by one or more assessment items as shown in the middle column. The right column provides the page numbers for the lessons in this book that reteach the Cluster B prerequisite skills

Skill Name	Assessment Items	Lesson Pages
Order Numbers to 20	5-6	76
Patterns	7-8	77
Compare Numbers to 20	9-10	78-79

Cluster B Challenge
Those students who demonstrate mastery of the skills on this page will not need to use the reteaching worksheets. Instead, these students can do the Cluster B Challenge found on page 81.

Grade 1, Chapter 7, Cluster B **71D**

USING THE LESSON

Lesson Goal
- Count tens and ones and write numbers to 20

What the Student Needs to Know
- Match objects in one-to-one correspondence.
- Write and count numbers to 10.
- Identify the same number shown by different arrangement of objects.

Getting Started
- Give each student some counters and a tens frame.
- Say: *Put one counter in each space. Now let's count the counters in the frame together. 1-2-3-4-5-6-7-8-9-10. How many counters fit in the frame?* (10)
- Call on several volunteers to count the counters in their frames. Ask: *Do all the frames hold ten counters?* (yes) *So, whenever you see a frame, how many counters will it hold?* (10)

What Can I Do?
- Read the question and the response. Then read and discuss the example.
- Point to the tens frame at the top of the page. Ask: *How many counters are in this frame?* (10) *We write numbers by writing the numbers of tens and then writing the number of ones.*
- Count the tens first. *How many tens are there?* (1) *What belongs on the line by this word "ten"?* (1) *That's right because there is only 1 ten.* Model writing the 1 in the correct place.
- Direct students' attention to the next part of the example. Say: *I see 1 ten, but how many ones are there? Let's count these extra counters together: 1-2-3-4-5-6-7, ?* (8) *How many ones?* (8) *Where do we write the 8?* (beside the word "ones")
- *We have 1 ten 8 ones. To write the number of counters we write the 1 first and then the 8. What number is 1 ten 8 ones?* (18)

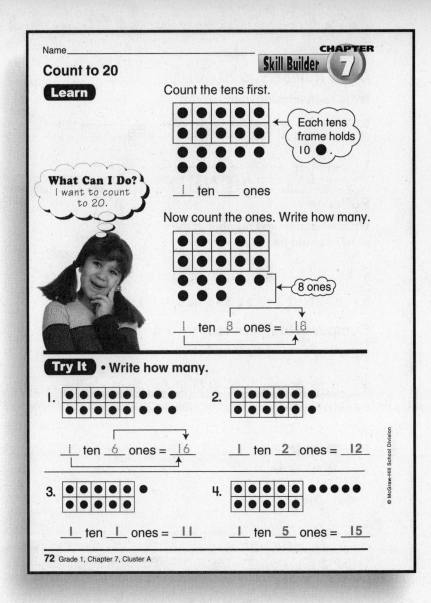

WHAT IF THE STUDENT CAN'T

Match with One-to-One Correspondence
- Have the student put one counter in each space of a tens frame. Then give him/her a set of cards on which you have drawn horizontal lines. Draw 1 to 9 lines per card.
- Tell the student to practice filling the frame and putting one counter on each line on the card.
- If the student had difficulty, cover half and see if the student can do that much. Based on your results, cover more or less of the frame.

Write and Count Numbers to 10
- Make flash cards for the numbers from 1–10. Also make cards with dots to represent the numbers from 1–10.
- Start with the dot cards. Have the student count the dots and say the number.
- When the student can go through those cards, spread out the number cards face up on the table. Have the student count the dots on the dot card and point to the matching number card.
- Have the student count the dots and write the number as he/she says it aloud.

Name_____

Power Practice • Write how many.

5. _1_ ten _3_ ones = _13_

6. _1_ ten _8_ ones = _18_

7. _1_ ten _0_ ones = _10_

8. _1_ ten _4_ ones = _14_

9. _2_ tens _0_ ones = _20_

10. _1_ ten _1_ one = _11_

11. _1_ ten _7_ ones = _17_

12. _1_ ten _9_ ones = _19_

Grade 1, Chapter 7, Cluster A **73**

USING THE LESSON

Try It

- Read the directions and make sure that students know what they are supposed to do.
- Work through the first exercise with students.
- Ask: *How many tens do you see?* (1 ten) *Where do we write the 1?* (on the line by the word "ten") *How many ones are there?* (6 ones) *What do we write by the word "ones"?* (6)
- *Now let's write the number. These arrows help us write the numbers in the right places in the answer. We write the tens first, so how many tens?* (1) *We write the ones next, so how many ones?* (6) *What number did you write?* (16)
- Have students complete exercises 2–4. Go over students' work with them.
- Some students who need more support counting to 20 may find it helpful to continue using a tens frame and counters.

Power Practice

- Read the directions with students and make sure they understand what to do.
- Have students complete the exercises. Then select a few exercises and have volunteers demonstrate counting tens and ones and writing the answers.
- Students who need more support counting to 20 may benefit from having tens frames and counters available.

Learn with Partners & Parents

- Have students make their own tens frames. Give each student an egg carton. Have them cut off two egg cups.
- Have students gather 20 counters, such as pennies, buttons, or pieces of macaroni. Then have them write the numbers from 10–19 on slips of paper and put them in a bag.
- Tell students to practice modeling the numbers on the papers using their tens frames and counters.

WHAT IF THE STUDENT CAN'T

Identify the Same Number of Objects

- Give the student some counters. Tell the student to model the number 10 with the counters.
- Then give the student a tens frame. Have the student model ten with the tens frame. Ask how the numbers compare. (They are the same.)
- Repeat the activity with other numbers from 10–19.

Complete the Power Practice

- Discuss each incorrect answer. Discuss a strategy the student can use to solve each one. Make tens frames and counters available if they will help.
- Then have the student redo each incorrect response and write the correct answer on its line.

Grade 1, Chapter 7, Cluster A **73**

USING THE LESSON

Lesson Goal
- Identify different arrangements of objects as representing the same number.

What the Student Needs to Know
- Identify same and different shapes.
- Match with one-to-one correspondence.

Getting Started
- Draw a tens frame on the chalkboard. Say: *Count with me as I draw a counter in each space: 1-2-3-4-5-6-7-8-9-10. How many counters?* (10)
- Draw another tens frame but orient it vertically. Repeat the activity. Ask: *What do you notice about the two tens frames?* (They show the same number.)

What Can I Do?
- Read the question and the response. Then read and discuss the example.
- Ask: *Look at the first tens frame. How many tens are there?* (1) *How many ones?* (2) *What number is 1 ten 2 ones?* (12)
- Now count the cherries. *How many are there?* (12)
- *Look at the next tens frame. How many tens are there?* (1) *How many ones?* (1) *What number is 1 ten 1 one?* (11)
- Write the numbers. *Which numbers match?* (the 12s)

Try It
- Read the directions and make sure that students know what they are supposed to do. Work through the first exercise. Ask: *How many tens and ones in the first set?* (1 ten 5 ones) *What number is that?* (15)
- Count each set of cherries. *How many are in each?* (10; 15) *Which numbers match?* (the 15s)

Power Practice
- Have students complete the practice items. Then review each answer.

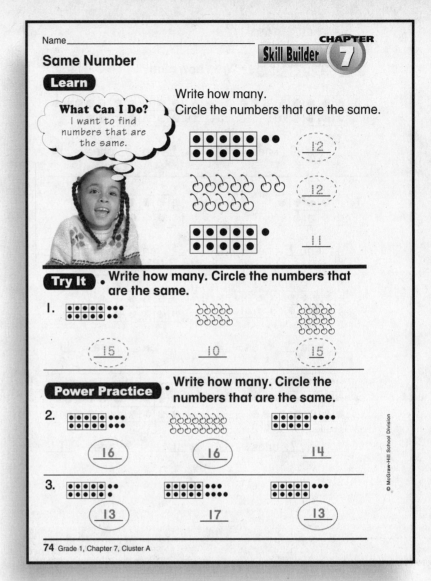

WHAT IF THE STUDENT CAN'T

Identify Same and Different Shapes
- Make a set of cards with shapes on them. Make an even number of circles, squares, and triangles. Then have the student use the cards to play concentration.

Match with One-to-One Correspondence
- Draw a tens frame on an index card. On other cards draw circles to show numbers 1–9.
- Have the student practice putting counters on the circles on the 1–9 cards. Then have him/her put counters on the tens frame and a 1–9 card.

Complete the Power Practice
- Discuss each incorrect answer. Have the student write the number of tens and ones if needed.

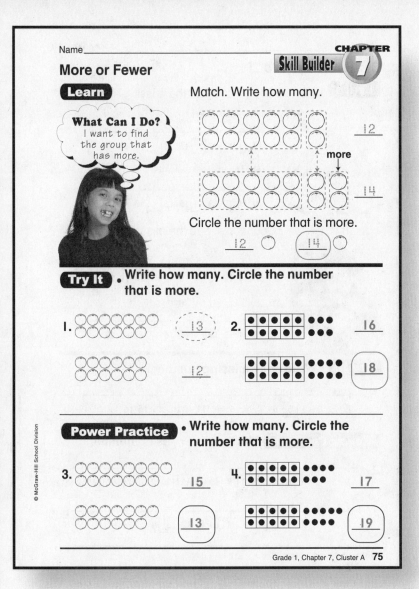

USING THE LESSON

Lesson Goal
- Determine which group has more.

What the Student Needs to Know
- Read, write, and count numbers to 20.
- Match objects in one-to-one correspondence.

Getting Started
- Say: *You can choose an apple or an orange. Who wants an apple?* Record the number. Ask: *Who wants an orange?* Record the number.
- Ask: *Do more students want apples or oranges?* Discuss students' answers.

What Can I Do?
- Read the question and the response. Then read and discuss the example.
- Say: *Let's count tens and ones to find how many oranges are in each group. Draw a box around groups of ten. Then count the ones. How many oranges are in the top group?* (12) *In the bottom group?* (14)
- *You matched the tens in the two groups. How many more oranges can you match?* (2 more in each group.) *Draw boxes around the ones that match. Which number is more?* (14)

Try It
- Read the directions and make sure that students know what they are supposed to do. Work through the first exercise. Ask: *How many tens and ones are in the first group?* (1 ten 3 ones) *What number is that?* (13)
- *How many tens and ones are in the next group?* (1 ten 2 ones) *What number is that?* (12) *Which number is more?* (13) *Circle that number.*

Power Practice
- Have students complete the practice items. Then review each answer.

Grade 1, Chapter 7, Cluster A

USING THE LESSON

Lesson Goal
- Order whole numbers through 20.

What the Student Needs to Know
- Read, write, and count to 20.
- Identify what comes before and after.

Getting Started
- Draw on the chalkboard a number line with 11 tick marks. Write *10* below the first mark. Do not write the other numbers.
- Say: *What number should I write next after 10?* (11) Have a volunteer write the number on the line. Repeat until students complete a 10–20 number line.

What Can I Do?
- Read the question and the response. Then read and discuss the example.
- Say: *This number line is missing some numbers. We can count on to find them. Put a finger on 10 and count on with me to find the first missing number: 10-11-12-13-14. What number is missing?* (14)
- Repeat the process to find the next missing number. Then have students count the numbers on the line from 10–20.

Try It
- Read the directions and make sure that students know what they are supposed to do.
- Work through the first exercise with students.
- Ask: *How can we find the first missing number?* (count on) *Let's count together: 1-2-3-4-__?* (5) *What number goes in the first box?* (5)
- Repeat the process for the rest of the missing numbers. Have students count the numbers on the line to check their work.

Power Practice
- Have students complete the practice items. Then review each answer.

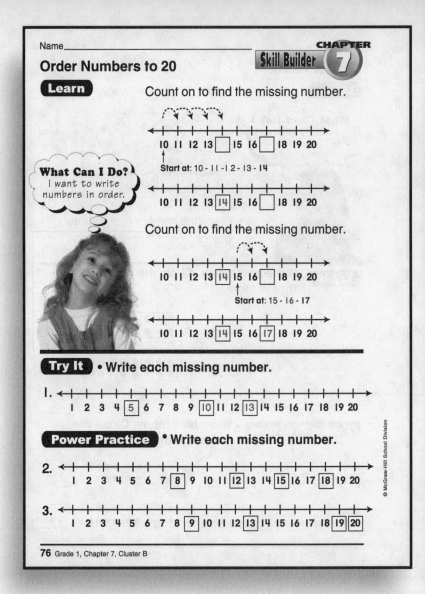

WHAT IF THE STUDENT CAN'T

Read, Write, and Count Numbers to 20
- Have students make a physical number line. Give the student a length of string with a large paper clip tied to one end, 20 pieces of elbow macaroni, and a marker. Have the student write the numbers 1–20 on the macaroni and put the pieces on the string in order. Tie a paper clip at the end.
- Then have the student count to various numbers on the necklace and write them. The student can point to a number and name it or represent it with counters.

Identify What Comes Before and After
- Write *Start* on an index card and place it on the left side of the student's desk. Put 5 objects after the card.
- Say: *This line starts here.* Point to the *Start* card. Then point to the second object. Ask: *Which thing comes before this one in line? Which thing comes just after this one?* Continue by pointing to other objects.

Complete the Power Practice
- Discuss each incorrect answer. Refer students to the completed number line at the top of the page if needed.

Name_____

Patterns

Skill Builder — CHAPTER 7

Learn

What Can I Do?
I want to find the next number in a pattern.

2 4 6 8 2 4 6 8 2 4 6 8 2 ___

Look for chunks.

[2 4 6 8] [2 4 6 8] [2 4 6 8] 2 ___

Look for a pattern.

[2 4 6 8] [2 4 6 8] [2 4 6 8] 2 ___
↑
(2 4 6 8 repeats)

Write the number that could come next.

[2 4 6 8] [2 4 6 8] [2 4 6 8] 2 **4**

Try It
Write what the next number in the pattern could be.

1. [3 1 3 2] [3 1 3 2] [3 1 3 2] 3 1 3 **2**

Power Practice
Write what the next number in the pattern could be.

2. 4 5 4 5 4 5 4 5 4 **5**

3. 5 6 7 5 6 7 5 6 7 5 6 **7**

4. 11 12 11 12 11 12 11 12 11 12 11 **12**

Grade 1, Chapter 7, Cluster B **77**

WHAT IF THE STUDENT CAN'T

Read, Write, and Count Numbers to 12
- Give the student an inch ruler. Have the student count the numbers, say them and point to each one.
- Tell the student to copy the numbers from the ruler. Say numbers at random and have the student write them.

Identify the Repeating Elements in a Pattern
- Show the student how to make a 2-shape pattern. Push the blocks to separate the first chunk. Have the student separate the rest. Repeat with other patterns.

Identify the Next Shape in a Repeating Geometric Pattern
- Model a 2-shape pattern with pattern blocks. Have the student separate the pattern into chunks and name the repeating chunk. Give the student one of each of the pattern shapes. Have the student read the last chunk, place the missing shape in the pattern, and read the whole pattern to check.

Complete the Power Practice
- Discuss each incorrect answer. Have students circle chunks if needed.

USING THE LESSON

Lesson Goal
- Determine the next number in a repeating pattern.

What the Student Needs to Know
- Read and write numbers to 12.
- Identify the repeating elements in a pattern.
- Identify the next shape in a repeating geometric pattern.

Getting Started
- Draw a row of 4 rectangles on the chalkboard. In the first rectangle write the numbers 1234. Say: *I am going to make a pattern. I am going to repeat these numbers over and over.* Write 1234 in the second box. Ask: *What numbers should I write in the third box?* (1234) Repeat with the fourth box. Have students read the whole pattern.

What Can I Do?
- Read the question and the response. Then read and discuss the example.
- Say: *Let's read this pattern together.* Read the pattern with students. Say: *I can see that the numbers 2468 make a chunk. Let's find more chunks.* Model marking the chunks. Ask: *What pattern do you notice?* (2468 repeats) *I know that the pattern is 2468. I can find the missing number in the last chunk. The first number in the chunk is 2. What comes next?* (4)

Try It
- Read the directions and make sure that students know what they are supposed to do. Work through the first exercise.
- Ask: *What chunks can you find?* (3132) *Mark them. What pattern do you notice?* (3132 repeats) *Read the last chunk. What number comes next?* (2)

Power Practice
- Have students complete the practice items. Then review each answer.

Grade 1, Chapter 7, Cluster B **77**

USING THE LESSON

Lesson Goal
- Compare numbers to 20.

What the Student Needs to Know
- Read, write, and count numbers to 20.
- Identify the group with more or fewer.
- Identify what comes before and after.

Getting Started
- Draw 4 apples on the chalkboard. Write a 4 below them. Say: *I need more than 4 apples. Should I draw an apple or erase one?* (draw) Draw the apple. Ask: *How many apples are there now?* (5) *So 5 apples is more than 4 apples.* Repeat with more numbers.
- Say: *Now I need fewer apples. Do I draw one or erase one?* (erase) Continue as above to find fewer apples.

What Can I Do?
- Read the question and the response. Then read and discuss the example.
- Direct students' attention to the apples at the top of the page. Say: *To find the group that has more let's match as many apples as we can. First put a box around 4 apples in each group.*
- *I can see that the group of 6 has 2 apples that are not matched. So, 6 is more than 4.*
- Write > and < on the chalkboard. Point to the > symbol. Say: *This symbol tells us that the number on the left is more. You can remember that because the bigger side of the symbol is on the side with the bigger number. We say, "is greater than" for this symbol.* Write 6 > 4. Say: *Let's read this together: 6 is greater than 4.*
- Point to the < symbol. Say: *This symbol means "is less than." We can switch the numbers and turn the symbol the other way.* Write 4 < 6 on the chalkboard. Say: *We read this as 4 is less than 6.*

78 Grade 1, Chapter 7, Cluster B

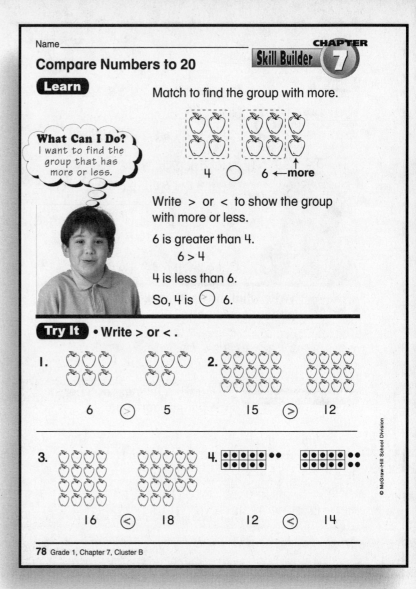

WHAT IF THE STUDENT CAN'T

Read, Write, and Count Numbers to 20
- Make flashcards for 1–20. Place the 1–10 cards in order on a chalkledge. Have the student read each card as you point to it. Then have the student point to each, read it, and turn it over.
- Then show one number. Have the student name it. Repeat for other numbers. Use the same process for 11–20.
- Mix up the cards. Have the student pick 5 cards without looking at them and give them to you. Read the cards and have the student write the numbers. Repeat several times.

Identify the Group with More or Fewer
- Use a set of dominoes or make a set of domino cards. Have the student compare the number of dots in the two sides of each domino and name the side that has more.
- Then have the student make domino cards. Each side should have between 10 and 20 dots. Mix up the cards. Have the student name the side that has fewer dots.

Name_____

Power Practice • Write > or < .

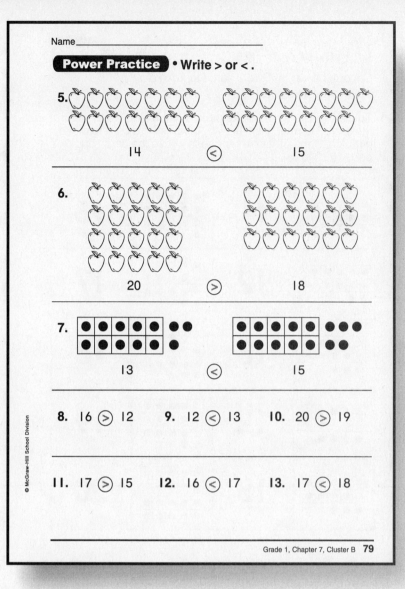

5. 14 < 15
6. 20 > 18
7. 13 < 15
8. 16 > 12
9. 12 < 13
10. 20 > 19
11. 17 > 15
12. 16 < 17
13. 17 < 18

WHAT IF THE STUDENT CAN'T

Identify What Comes Before and After

- Have the student open a book to page 10. Ask the student what page comes before page 10. Have the student check his/her work by flipping back a page.
- Then ask the student to name a page that comes after page 11. Have the student check his/her work by finding page 11 and the page he/she named.
- Continue with several other pages.

Complete the Power Practice

- Discuss each incorrect answer. Discuss a strategy the student can use to solve each one, such as matching and using a number line for exercises 5–7, and visualizing, modeling, drawing pictures, and using a number line for exercises 8–13.
- Make tens frames and counters available for modeling if they will help a student who needs support.
- Have the student redo each incorrect response and write the correct symbol in the circle.

USING THE LESSON

Try It

- Read the directions and make sure that students know what they are supposed to do.
- Work through the first exercise with students.
- Ask: *How many apples are in the first group?* (6) *How many apples are in the second group?* (5) *How can we find out which has more?* (Match the apples.) *Which group has more?* (6) *Which symbol do we write to show that 6 is greater than 5?* (>) *How do we know that we have written the correct symbol?* (The bigger side faces the bigger number.)
- Have students complete exercises 2–4. Go over students' work with them. Students who need more support may benefit from having manipulatives available.

Power Practice

- Read the directions with students. Make sure they understand what to do.
- Have students complete the exercises. Then select a few exercises and have volunteers demonstrate how they completed each comparison and how they decided which symbol to use. Have students read the comparisons.
- Point out that exercises 8–13 do not have pictures. Discuss strategies students can use to complete each comparison, such as modeling, visualizing, drawing pictures, and using a number line.

CHALLENGE

Lesson Goal
- Read, write, and count numbers through 20.

Introducing the Challenge
- Ask students if they have ever played a game with cards called Go Fish. Tell students that they are going to play a similar game. This time they will fish to match numbers 11–20.

Using the Challenge
- Read the directions with students.
- Then have each pair cut out a set of cards and mix them up. Have the students give 3 cards to each player and put the rest face down in a pile.
- Remind students to take turns. Explain that they have to match a number card with a card that shows a tens frame and counters for the same number.
- Tell students that if they start out with a match in their hands, they can put it down when it is their turn.
- Tell students that when they can make a match, they should put the two matching cards down in front of them.
- If a player runs out of cards, he/she can take the next card from the top of the pile for his/her next turn.
- Have students play the game.

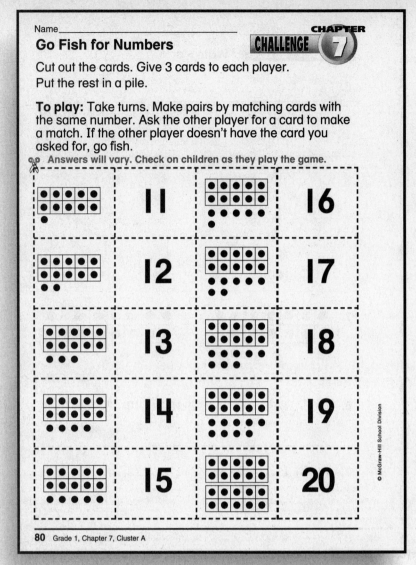

80   Grade 1, Chapter 7, Cluster A

Name_____

Secret Number Code

Write the numbers in each box in order.
Write the numbers from least to greatest.

1. | 8 | 10 |
    | 11 | 9 |

    _8_ _9_ (10) _11_

2. | 16 | 17 |
    | 15 | 14 |

    _14_ _15_ (16) _17_

3. | 13 | 11 |
    | 10 | 12 |

    _10_ _11_ (12) _13_

4. | 20 | 17 |
    | 19 | 18 |

    _17_ _18_ (19) _20_

5. | 15 | 10 |
    | 5 | 20 |

    _5_ _10_ (15) _20_

6. | 6 | 2 |
    | 8 | 4 |

    _2_ _4_ (6) _8_

| 3 = S |
| 4 = R |
| 6 = Y |
| 9 = A |
| 10 = T |
| 11 = P |
| 12 = E |
| 14 = O |
| 15 = T |
| 16 = W |
| 19 = N |
| 20 = F |

Match the numbers in the circles to the letters in the box.
Write the letters on the lines below. Find the secret number.

T W E N T Y
1. 2. 3. 4. 5. 6.

Grade 1, Chapter 7, Cluster B **81**

CHALLENGE

Lesson Goal
- Order numbers through 20.

Introducing the Challenge
- Ask students if they have ever solved a secret code. Explain that they are going to solve a puzzle to find a secret number. All they have to do is put some words in order to get the key for solving the code.

Using the Challenge
- Read the directions with students.
- Work through the first exercise with students.
- Point to the four numbers in the box. Explain that these numbers are all mixed up. They need to find the one that is the least—the one that would come first when they count. Ask students which number in the box is least. (8) Ask: *Which of the numbers in the box would come next?* (9) Continue with the other two numbers.
- Explain that for some exercises the numbers won't count on by ones. They will have to think about the order of all the numbers.
- Point out the circles around some numbers. Explain that the numbers that they write on those lines are part of the secret code. Ask students which number that they wrote in the first example is in a circle. (10)
- Direct students' attention to the code box on the right. Have students find 10 in the code box. Ask which letter matches with 10. (T) Point out the lines at the bottom of the page. Explain that the students should write *T* on the first line. To find the rest of the letters for the secret number, they have to find the rest of the secret numbers.
- Have students complete exercises 2–6 and solve the secret code.

Grade 1, Chapter 7, Cluster B **81**

Name_____

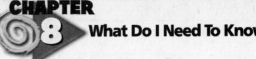

Skip Count by 5s and 10s

Skip count by fives.
Write each missing number.

1. 0   5   10   ☐   20   ☐   30   35   ☐

Skip count by tens.
Write each missing number.

2. 0   10   20   ☐   40   50   ☐   70   ☐

Number Patterns

Write each number.

1	2	3	4	5	6	7	8	9	10
11	12	13	14	15	16	17	18	19	20
21	22	23	24	25	26	27	28	29	30
31	32	33	34	35	36	37	38	39	40

3. 1 more than 25 _____   **4.** 10 more than 25 _____

Compare Numbers to 100

Write how many.
Write >, <, or =.

5.

_____   _____   _____

81A Use with Grade 1, Chapter 8, Cluster A

Name_____

Add Sums to 12

Add. Write each sum.

6.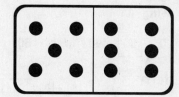

7.   4
   + 5
   ___

5 + 6 = _____

Subtract from 12

Subtract. Write each difference.

8.

9. 12 − 9 = _____

10 − 4 = _____

10.   9
    − 3
    ___

Use with Grade 1, Chapter 8, Cluster B **81B**

CHAPTER 8 PRE-CHAPTER ASSESSMENT

Assessment Goal
This two-page assessment covers skills identified as necessary for success in Chapter 8 Money. The first page assesses the major prerequisite skills for Cluster A. The second page assesses the major prerequisite skills for Cluster B. When the Cluster A and Cluster B prerequisite skills overlap, the skill(s) will be covered in only one section.

Getting Started
- Allow students time to look over the two pages of the assessment. Point out the labels that identify the skills covered.
- Have students find math vocabulary terms used in the assessment. List vocabulary terms on the board as students identify them. If necessary, review the meanings of all essential math vocabulary.

Introducing the Assessment
- Explain to students that these pages will help you know if they are ready to start a new chapter in their math textbooks.
- Students who have transferred from another school may not have been introduced to some of these skills. Encourage students to do their best and assure them you will help them learn any needed skills.

Cluster A Challenge
Those students who demonstrate mastery of the skills on this page will not need to use the reteaching worksheets. Instead, these students can do the Cluster A Challenge found on pages 92-93.

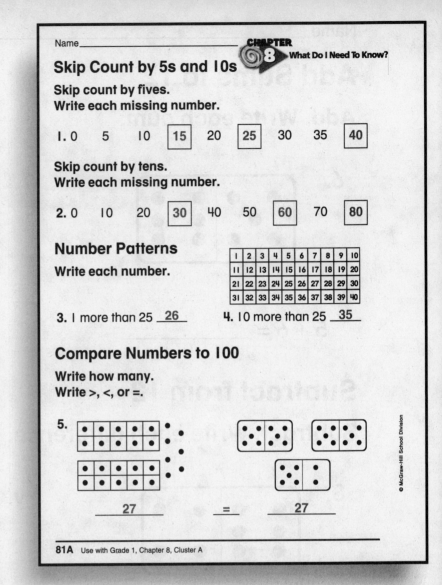

CLUSTER A PREREQUISITE SKILLS

The skills listed in this chart are those identified as major prerequisite skills for students' success in the lessons in Cluster A of the chapter. Each skill is covered by one or more assessment items as shown in the middle column. The right column provides the page numbers for the lessons in this book that reteach the Cluster A prerequisite skills.

Skill Name	Assessment Items	Lesson Pages
Skip Count by 5s	1	82
Skip Count by 10s	2	83
Number Patterns: 1 More	3	84
Number Patterns: 10 More	4	85
Compare Numbers to 100	5	86-87

Name_____

Add Sums to 12

Add. Write each sum.

6.

7. $\begin{array}{r} 4 \\ +5 \\ \hline 9 \end{array}$

$5 + 6 = \underline{\quad 11 \quad}$

Subtract from 12

Subtract. Write each difference.

8.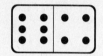

9. $12 - 9 = \underline{\quad 3 \quad}$

$10 - 4 = \underline{\quad 6 \quad}$

10. $\begin{array}{r} 9 \\ -3 \\ \hline 6 \end{array}$

Use with Grade 1, Chapter 8, Cluster B **81B**

CHAPTER 8 PRE-CHAPTER ASSESSMENT

Alternative Assessment Strategies

- Oral administration of the assessment is appropriate for younger students or those whose native language is not English. Read the skills title and directions one section at a time. Check students' understanding by asking them to tell you how they will do the first exercise in the group.
- For some skill types you may wish to use group administration. In this technique, a small group or pair of students complete the assessment together. Through their discussion, you will be able to decide if supplementary reteaching materials are needed.

Intervention Materials

If students are not successful with the prerequisite skills assessed on these pages, reteaching lessons have been created to help them make the transition into the chapter.

Item correlation charts showing the skills lessons suitable for reteaching the prerequisite skills are found beneath the reproductions of each page of the assessment.

CLUSTER B PREREQUISITE SKILLS

The skills listed in this chart are those identified as major prerequisite skills for students' success in the lessons in Cluster B of the chapter. Each skill is covered by one or more assessment items as shown in the middle column. The right column provides the page numbers for the lessons in this book that reteach the Cluster B prerequisite skills.

Skill Name	Assessment Items	Lesson Pages
Add Sums to 12	6-7	88-89
Subtract from 12	8-10	90-91

Cluster B Challenge

Those students who demonstrate mastery of the skills on this page will not need to use the reteaching worksheets. Instead, these students can do the Cluster B Challenge found on pages 94-95.

Grade 1, Chapter 8, Cluster B **81D**

USING THE LESSON

Lesson Goal
- Count by fives to 100.

What the Student Needs to Know
- Read, write, and count numbers to 100.
- Identify number patterns.
- Extend number patterns.

Getting Started
- Give each student a card. Tell students to form a line. Have them count off and write the number they say on their cards.
- Point to the student with the 2 card. Say: *Step forward and say your number.* Repeat for students holding the 4, 6, 8, 10, and 12 cards. Ask: *What number do I want next?* (14) *What pattern do you see?* (counting by twos) Have students continue counting by twos. Repeat the process to skip count by fives.

What Can I Do?
- Read the question and the response. Then read and discuss the example.
- Say: *We need to skip count by fives, Look at the chart. Let's count together by fives: 5-10-15-20-25-30. What pattern do you notice?* (Every number has a ones digit that is 5 or 0.)
- *Let's count to find the first missing number: 5-10-__?__.* (15) *Count on for the next missing number: 20-25-__?__.* (30)

Try It
- Work through the first exercise with students.
- Say: *Let's count to find the first missing number: 5-10-15-__?__.* (20) *Count on to find the second missing number: 25-30-__?__.* (35).
- Have students complete exercise 2. Check their work.

Power Practice
- Have students complete the practice items. Then review each answer.

WHAT IF THE STUDENT CAN'T

Read, Write, and Count Numbers to 100
- Give the student a hundred chart and some tens models. Count aloud with the student to 20, pointing to the numbers on the chart as you count. Model 20 with the tens rods. Add some ones cubes. Have the student count and write the number. Continue with other numbers.

Identify Number Patterns
- Write a two-number pattern on the chalkboard, such as 25252525. Tell students that this pattern has two numbers that repeat. Have students circle the repeating chunk. Repeat for other 2- to 4-number repeating patterns.

Extend Number Patterns
- Write a two-number repeating pattern on the chalkboard. Have the student circle the chunk that repeats each time. Then have the student name the number that comes next. Repeat with other patterns.

Complete the Power Practice
- Discuss each incorrect answer. Have the student use a hundred chart if necessary. Have the stuent correct each pattern and say it aloud.

Name_____

Skip Count by 10s

Skill Builder — CHAPTER 8

Learn

What Can I Do? I want to skip count by tens.

Skip count by tens. Write each missing number.

10 20 30 ___ ___

Look for a pattern.

1	2	3	4	5	6	7	8	9	10
11	12	13	14	15	16	17	18	19	20
21	22	23	24	25	26	27	28	29	30
31	32	33	34	35	36	37	38	39	40
41	42	43	44	45	46	47	48	49	50

10 20 30 _40_ _50_

Try It

Skip count by tens. Write each missing number.

1. 10 20 _30_ 40 _50_

2. 50 60 _70_ _80_ 90

Power Practice

Skip count by tens. Write each missing number.

3. 10 _20_ 30 40 50 _60_

4. 10 20 _30_ _40_ _50_ _60_ 70 _80_ _90_ _100_

Grade 1, Chapter 8, Cluster A **83**

WHAT IF THE STUDENT CAN'T

Read, Write, and Count Numbers to 100

- Give the student a hundred-chart. Fold it so that only 1–30 is visible. Have the student count the numbers. Call out numbers randomly and have the student point to them. Then say several numbers and have the student write them.

Identify Number Patterns

- Draw 5 boxes in a row. Write two numbers in the first box. Have the student write the two numbers in the other boxes and read the pattern. Write other 2-number patterns on the chalkboard. Have the student circle the chunks.

Complete the Power Practice

- Discuss each incorrect answer. Have the student use a hundred chart if necessary. Have the student write the correct numbers for each pattern and say the pattern aloud.

USING THE LESSON

Lesson Goal
- Count by tens to 100.

What the Student Needs to Know
- Read, write, and count numbers to 100.
- Identify number patterns.

Getting Started
- Give each student a card. On each card write a number from 5–100, skip counting by fives. Have students stand in a line. Tell them to arrange themselves in order. Then have students count by fives, showing their cards as they say their numbers.
- Say: *If you have a zero in your number step forward. Let's count these cards together: 10, 20, 30, . . . 100.*

What Can I Do?
- Read the question and the response. Then read and discuss the example.
- Say: *We need to skip count by tens to find the missing numbers. Look at the chart. Let's count by tens: 10-20-30-40-50. What pattern do you notice?* (Each number has a ones digit that is 0.)
- *Let's count to find the missing numbers: 10-20-30- ? .* (40, 50) *If we continue counting, what would be the next numbers?* (60-70-80-90-100)

Try It
- Read the directions. Make sure students know what to do.
- Work through the first exercise.
- Say: *Let's count to find the first missing number: 10-20- ? .* (30) *Count on to find the second missing number: 40- ? .* (50)
- Have students complete exercise 2. Check their work.

Power Practice
- Have students complete the practice items. Then review each answer.

Grade 1, Chapter 8, Cluster A **83**

USING THE LESSON

Lesson Goal
- Identify the number that is 1 more than a given number.

What the Student Needs to Know
- Read, write, and count numbers to 100.
- Identify number patterns.
- Extend number patterns.

Getting Started
- Tell students that you are going to say a number and then toss a beanbag to a student. That student will begin counting on until you tell him or her to stop. Then you will say another number, and have the student toss the beanbag to another student, who then counts on from that number. Do the activity several times.

What Can I Do?
- Read the question and the response. Then read and discuss the example.
- Say: *You know how to count on. You count on to find one more than a number. We need to find 1 more than 20. Find 20 on the chart. Follow the arrow from 20 to the number that is 1 more. What number is 1 more than 20?* (21)

Try It
- Read the directions and make sure students know what they are supposed to do.
- Work through the first exercise with students.
- Say: *Count on to find one more than 30. Count with me: 30- ? .* (31) *So what number is one more than 30?* (31)
- Have students complete exercise 2. Check their work.

Power Practice
- Have students complete the practice items. Then review each answer.

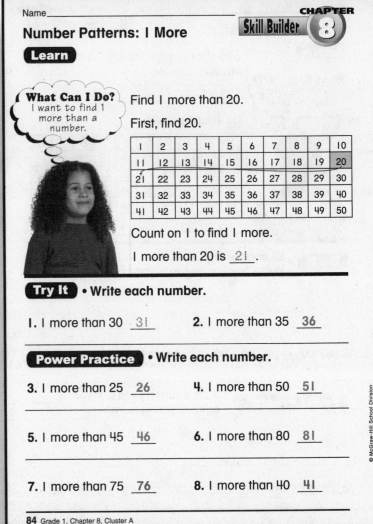

Number Patterns: 1 More

Learn

What Can I Do? I want to find 1 more than a number.

Find 1 more than 20.

First, find 20.

1	2	3	4	5	6	7	8	9	10
11	12	13	14	15	16	17	18	19	20
21	22	23	24	25	26	27	28	29	30
31	32	33	34	35	36	37	38	39	40
41	42	43	44	45	46	47	48	49	50

Count on 1 to find 1 more.

1 more than 20 is 21 .

Try It • Write each number.

1. 1 more than 30 31
2. 1 more than 35 36

Power Practice • Write each number.

3. 1 more than 25 26
4. 1 more than 50 51
5. 1 more than 45 46
6. 1 more than 80 81
7. 1 more than 75 76
8. 1 more than 40 41

WHAT IF THE STUDENT CAN'T

Read, Write, and Count Numbers to 100
- Use a hundred chart and count to 100 together. Give the student a bingo card on which you have written numbers. Write those numbers on slips of paper and put them in a bag. Play bingo. Repeat, but switch roles. Have the student prepare the materials and call the numbers.

Identify Number Patterns
- Give the student a hundred chart. Highlight a column. Have the student identify the pattern. Repeat with another column. Have the student compare the two patterns.

Extend Number Patterns
- Write a series of numbers on the chalkboard. The numbers should count on by ones, twos, fives, or tens. Have the student identify the pattern, and name the next number in the pattern. Repeat for other patterns.

Complete the Power Practice
- Discuss each incorrect answer. Have the student use a hundred chart if necessary. Have the student write the correct number that is 1 more.

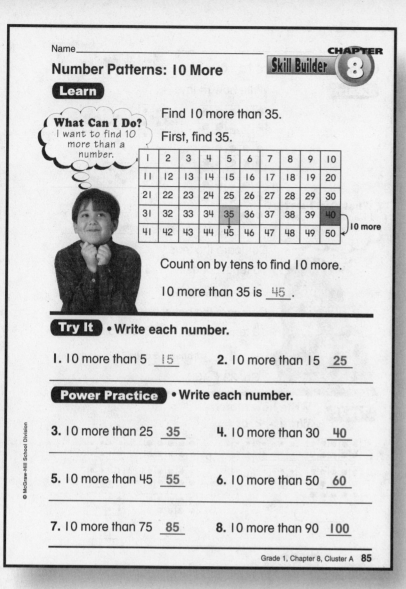

Number Patterns: 10 More
Skill Builder — Chapter 8

Learn

What Can I Do? I want to find 10 more than a number.

Find 10 more than 35.

First, find 35.

Count on by tens to find 10 more.

10 more than 35 is __45__.

Try It • Write each number.

1. 10 more than 5   __15__
2. 10 more than 15 __25__

Power Practice • Write each number.

3. 10 more than 25 __35__
4. 10 more than 30 __40__
5. 10 more than 45 __55__
6. 10 more than 50 __60__
7. 10 more than 75 __85__
8. 10 more than 90 __100__

Using the Lesson

Lesson Goal
- Identify the number that is 10 more than a given number.

What the Student Needs to Know
- Read, write, and count numbers to 100.
- Identify number patterns.
- Extend number patterns.

Getting Started
- Have students skip count by tens with you from 10 to 100. Ask several volunteers to write the numbers on the chalkboard. Discuss the pattern as you count on one ten more each time.

What Can I Do?
- Read the question and the response. Then read and discuss the example. Say:
- *You know how to skip count by tens. Look at the number chart. Put your finger on 10. Count down the column with me: 10-20-30-40-50. How much more is each number than the number above it?* (10 more)
- *Now put your finger on 5. Count down this column with me: 5-15-25-35-45. How much more is each number than the number above it?* (10 more) *What pattern do you notice?* (The tens go up by one, but the ones stay the same.)
- *Find 35. What number is 10 more than 35?* (45)

Try It
- Read the directions. Make sure students know what to do.
- Work through the first exercise.
- Say: *What number do you start with?* (5) *How do you find 10 more than that?* (Look directly below it on the chart.) *What number is 10 more than 5?* (15)
- Have students complete exercise 2. Check their work.

Power Practice
- Have students complete the practice items. Then review each answer.

What If The Student Can't

Read, Write, and Count Numbers to 100
- Use a hundred chart and count to 100 together. Give the student a book that has about 100 pages. Have the student open the book randomly to a page. Have the student read the page number. Then have the student read the numbers on the next few pages. Say a number and have the student write it and find that page in the book.

Identify Number Patterns
- Give the student a hundred chart. Highlight a column. Have the student identify the pattern. Repeat with another column. Have the student compare the two patterns.

Extend Number Patterns
- Write a three-number repeating pattern on the chalkboard. Use numbers that count on by ten, such as 35, 45, 55, 35, 45, 55, … Have the student identify the pattern and name the next number. Repeat for other patterns.

Complete the Power Practice
- Discuss each incorrect answer. Have the student use a hundred chart if necessary. Have the student write the number that is 10 more.

USING THE LESSON

Lesson Goal
- Compare numbers to 100.

What the Student Needs to Know
- Read, write, and count numbers to 100.
- Identify the group with more or fewer.
- Identify what comes before and after.

Getting Started
- Have students work with partners. Give each student 10 counters and 10 slips of paper. Have them write the numbers 1–10 on the slips of paper. Give each pair an index card with > on the front and = on the back.
- Say: *Turn your papers facedown. Each partner takes a paper. Compare your numbers.* Have students arrange their numbers and one symbol to make a true statement. Tell students they can turn the > symbol around to show <.
- Have students check their work with counters. Then have them read their statements.

What Can I Do?
- Read the question and the response. Then read and discuss the example.
- Point to the tens frames. Ask: *How many tens are there?* (3) *How many ones?* (2) *What number is 3 tens 2 ones?* (32)
- Point to the top domino. Ask: *How many dots are on the dominoes that have 5 dots on each side?* (10) *So a full domino has 10 dots and can be called 1 ten. How many tens are there?* (3) *How many ones?* (5) *What is 3 tens 5 ones?* (35)
- *To decide which number is greater we compare the tens first. How many tens are there in 32?* (3) *How many in 35?* (3) *The number of tens is the same. So, we can't tell which number is greater yet.*
- *Now let's compare the ones. How many ones are there in 32?* (2) *in*

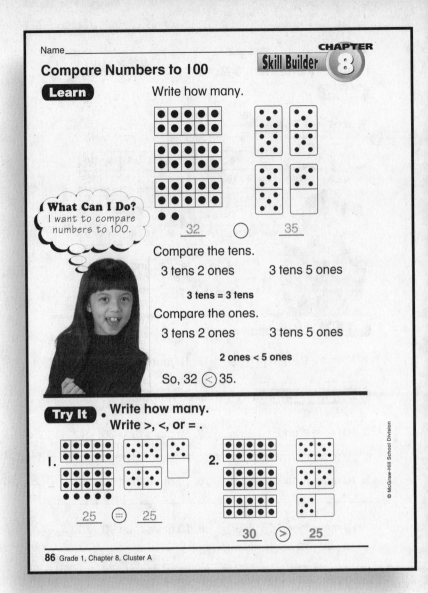

WHAT IF THE STUDENT CAN'T

Read, Write, and Count Numbers to 100
- Make sets of flashcards for the numbers from 1–20, 21–40, 41–60, 61–80, 81–100. Check the student's recognition of the numbers in each set. Remove any numbers the student can name without hesitation.
- Organize the troublesome numbers by tens. Give the student two numbers per day to study. Have the student keep copies of the numbers on his or her desk. Periodically during the day, ask the student to point to, name, and write the two numbers.

Identify the Group with More or Fewer
- Put about 40 small counters or other objects in a bag. Have the student reach in and take a handful. Tell the student to count the objects and write the number. Then have the student repeat the activity. Ask the student which group has more. Have the student check by matching the two groups of objects. Have the student circle the group that is more with a piece of yarn. Repeat several times. Then try the activity asking the student to identify the group that has fewer.

Name_____

Power Practice • Write how many.
Write >, <, or = .

3. 28 ⊂<⊃ 30   4. 31 ⊂>⊃ 30

5. 15 ⊂=⊃ 15   6. 26 ⊂<⊃ 27

Write >, <, or = .

7. 8 ⊂>⊃ 7   8. 10 ⊂=⊃ 10   9. 9 ⊂<⊃ 10

10. 10 ⊂<⊃ 11   11. 15 ⊂>⊃ 13   12. 18 ⊂>⊃ 17

13. 21 ⊂>⊃ 20   14. 25 ⊂<⊃ 28   15. 26 ⊂>⊃ 24

16. 30 ⊂>⊃ 29   17. 36 ⊂>⊃ 35   18. 40 ⊂=⊃ 40

Grade 1, Chapter 8, Cluster A **87**

WHAT IF THE STUDENT CAN'T

Identify What Comes Before and After

- Point to a number on a 0–20 number line. Ask the student to point to all the numbers that come before that number. Help the student see that all numbers to the left of that number come before the number. Have the student read all the numbers aloud. Repeat several times with a different number.
- Then follow the same procedure with numbers that come after a given number.
- Finally, repeat the activity with a hundred chart.

Complete the Power Practice

- Discuss each incorrect answer. Discuss a strategy the student can use to solve each one. Have tens frames and counters available.
- Then have the student redo each incorrect response and write the correct answer in the circle.

USING THE LESSON

35? (5) Compare 2 and 5. (2 < 5) Since 2 is less than 5, 32 is less than 35. How do we write that? (32 < 35)

Try It

- Read the directions and make sure students know what they are supposed to do.
- Work through the first exercise with students.
- Say: *Look at the tens frames. How many tens are there?* (2) *How many ones?* (5) *What is 2 tens 5 ones?* (25)
- *Now look at the dominoes. How many tens are there?* (2) *How many ones?* (5) *What is 2 tens 5 ones?* (25)
- Ask: *How do the tens compare?* (same) *How do the ones compare?* (same) *Both the tens and ones are the same. That means the numbers are ? * (equal) *What symbol do we write in the circle?* (=)
- Have students complete exercise 2. Go over students' work with them.

Power Practice

- Point out that exercises 7–18 do not have models. Discuss strategies students can use to determine the greater number, such as comparing tens and ones and visualizing.
- Have students complete the exercises. Then select a few exercises and have volunteers demonstrate comparing tens and ones and writing the answers.
- Students who need more support may benefit from having manipulatives available.

Learn with Partners & Parents

- Have students write > on an index card. A partner or family member writes two numbers on separate slips of paper. The student makes two statements for the number, for example, 24 > 14 and 14 < 28.

Grade 1, Chapter 8, Cluster A **87**

USING THE LESSON

Lesson Goal
- Add basic facts to 12.

What the Student Needs to Know
- Read, write, and count to 12.
- Identify more than or less than a number.
- Add with sums to 8.

Getting Started
- Ask two students to stand together with their backs to the chalkboard. Ask: *How many students do I have standing in this group?* (2) Write the number 2 on the chalkboard above the students' heads.
- Then have a group of 6 students also stand at the chalkboard a short distance from the other students. Ask: *How many students are in this group?* (6) Write the number above their heads.
- Write a + between the numbers. Say: *I want to add these two groups. What do I have to do?* (Put the groups together and count them) Have the groups move closer together. Say: *Let's count them together: 1-2-3-4-5-6-7-8. How many students altogether?* (8) Repeat with other groups.

What Can I Do?
- Read the question and the response. Then read and discuss the example.
- Point to the teddy bears in the first box. Ask: *How many teddy bears are in this box?* (4) *How many bears are in the other box?* (8) *What do we have to do?* (add) *We can put the groups together and count them.*
- Draw students' attention to the second group of boxes. Say: *These bears have numbers to show us how to count. Put a finger on each bear as we count together: 1-2-3-4-5-6-7-8-9-10-11-12. How many bears in all?* (12) *What is 4 + 8?* (12) *What do we write as the sum?* (12)

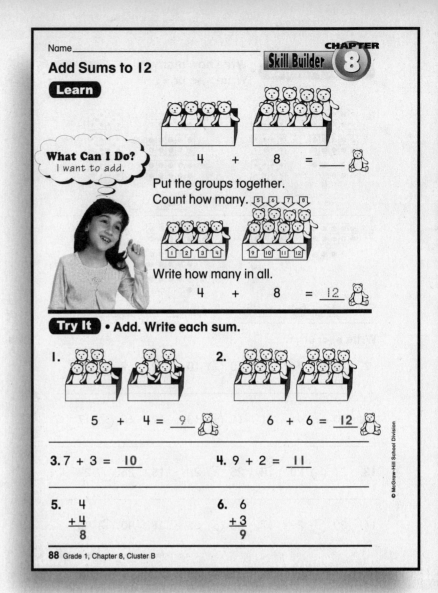

WHAT IF THE STUDENT CAN'T

Read, Write, and Count Numbers to 12
- Write the numbers 1–12 on sticky notes. Place them in order on the student's work area. Have the student read the numbers with you and also several times on his or her own.
- Mix the notes up. Say each number and have the student point to it.
- Then count with the student. Have the student find the number as he or she says it and place the sticky notes in order.
- Write the numbers 1–12 on the chalkboard. Have the student say each number as he or she traces it.

Identify More Than or Less Than a Number
- Make cube trains with 6, 4, and 8 connecting cubes. Point to the 6-cube train. Ask: *How many cubes are in this train?* (6) *Which of these trains has more than 6 cubes? How many cubes are in the train that has more?* (8) If students have difficulty, have them compare the train lengths. Repeat the activity, asking about fewer cubes. Then go on to use other numbers.

Name_____

Power Practice • Add. Write each sum.

7. 7 + 4 = 11
8. 5 + 3 = 8
9. 8 + 2 = 10
10. 6 + 4 = 10

11. 9 + 3 = 12
12. 3 + 6 = 9

13. 6 + 5 = 11
14. 4 + 3 = 7

15. 8 + 3 = 11
16. 7 + 5 = 12

17. 8
    +4
    ――
    12

18. 9
    +1
    ――
    10

19. 3
    +3
    ――
     6

20. 4
    +2
    ――
     6

Grade 1, Chapter 8, Cluster B 89

USING THE LESSON

Try It
- Read the directions and make sure that students know what they are supposed to do.
- Work through the first exercise with students.
- Ask: *How many bears are in the first box?* (5) *How many in the second box?* (4) *How can we find out how much 5 + 4 is?* (Put the groups together and count.) *Put your fingers on the bears. Let's count: 1-2-3-4-5-6-7-8- ?* (9) *How many bears are there? How much is 5 + 4?* (9) *What do we write as the sum?* (9)
- Have students complete exercises 2–6. Go over students' work with them.
- Point out that exercises 3–6 do not have pictures. Discuss strategies students can use to solve the additions, such as adding doubles, counting on, using related facts, modeling, visualizing, or drawing pictures.
- Remind students that some additions, such as exercises 5 and 6, are shown in columns instead of rows.
- Students who need more support adding may benefit from having manipulatives or number lines available.

Power Practice
- Read the directions with students and make sure they understand what to do.
- Point out that exercises 11–20 do not have pictures. Discuss strategies students can use to add if they don't remember a fact. They can add doubles, use related facts, count on, draw pictures, model, and visualize.
- Have students complete the exercises. Then select a few exercises and have volunteers demonstrate adding.
- Students who need more support may benefit from having manipulatives available.

WHAT IF THE STUDENT CAN'T

Add with Sums to 8
- Make a set of flashcards for facts with sums to 8. Use the cards with the student to find the most troublesome facts.
- Model a fact by placing counters next to the numbers on the card. Have the student add them and say the fact.
- Then have the student sing the fact to the tune of "The Farmer in the Dell." For example:

    "Oh, five and three is eight,
    Five and three is eight,
    Heigh-ho, the merry-o,
    five and three is eight."

- Repeat this activity with two or three facts each day.

Complete the Power Practice
- Discuss each incorrect answer. Discuss a strategy the student can use to solve each one. Have manipulatives available for the student.
- Then have the student redo each incorrect response and write the correct answer on the line.

Grade 1, Chapter 8, Cluster B **89**

USING THE LESSON

Lesson Goal
- Subtract numbers from 12.

What the Student Needs to Know
- Read, write, and count numbers to 12.
- Identify more than or less than a number.
- Add with sums to 12.

Getting Started
- Line up 10 books along the chalkledge. Ask a volunteer to count them. Then say: *I am going to take away three of these books.* Take the three books and write the subtraction problem on the chalkboard.
- Ask: *How many books are left?* (7) Write 7 as the difference.
- Model other subtraction problems with numbers to 12. Have students write the problems on the chalkboard.

What Can I Do?
- Read the question and the response. Then read and discuss the example.
- Point to the dolls on the shelves. Ask: *How many dolls are there?* (12) This little girl says she needs 3 dolls. We can subtract to find out how many are left. What subtraction problem do we need to solve? (12 – 3)
- One way to subtract is to cross out. Look at the dolls on the shelves below. We want to subtract 3. Let's count the dolls that are crossed out: 1-2-3. How many dolls are crossed out? (3) Now let's count the dolls that are left: 1-2-3-4-5-6-7-8-9. How many dolls are left? (9) So 12 – 3 = 9.

WHAT IF THE STUDENT CAN'T

Read, Write, and Count Numbers to 12
- In an egg carton, number the egg cups from 1–12. Have the student point to each number and read it several times.
- Give the student 12 counters. Tell the student to put 1 counter each in the number of egg cups you name. Demonstrate: Say "5 cups" and drop 1 counter in each of the 5 cups. Say each number as you come to it. Have the student complete the activity using other numbers. Then have the student write the number of counters placed in the cups each time.

Identify More Than or Less Than a Number
- Play a guessing game with the student. Give the student a 1–12 number line. Say: *I am thinking of a number. It is more than 5. It is less than 7. What number is it?* (6) Discuss why 6 is the correct answer. Have the student show you on the number line why 6 is correct. Repeat the activity several times.
- Then switch roles with the student. Have the student make up the clues and you guess the mystery number.

Name_____

Power Practice • Subtract. Write the difference.

7. 12 − 4 = __8__

8. 10 − 5 = __5__

9. 8 − 6 = __2__

10. 11 − 2 = __9__

11. 10 − 3 = __7__   12. 12 − 6 = __6__

13. 12 − 9 = __3__   14. 11 − 7 = __4__

15. 8 − 1 = __7__   16. 11 − 5 = __6__

17. 11
    −3
    ___
     8

18. 10
    −8
    ___
     2

19. 7
    −7
    ___
     0

20. 12
    −8
    ___
     4

Grade 1, Chapter 8, Cluster B **91**

WHAT IF THE STUDENT CAN'T

Add with Sums to 12

- Make a set of 24 cards with sums to 12. Create the cards in pairs so that there are two cards for each sum. For example, you can make the cards 4 + 8 and 6 + 6 for the sum of 12.
- First use the cards as flash cards. Check that the student is able to add each sum.
- Then mix up the cards and place them face down on the table. Play Concentration with the student. Have the student say each fact as he or she turns a card over. Match pairs with the same sum.

Complete the Power Practice

- Discuss each incorrect answer. Discuss a strategy the student can use to solve each problem. Have manipulatives available for students.
- Then have the student redo each incorrect response and write the correct answer on the line.

USING THE LESSON

Try It

- Read the directions and make sure students know what they are supposed to do.
- Work through the first exercise with students.
- Ask: *How many dolls are on these shelves?* (10) *How many do we subtract?* (4) *What can we do to show the subtraction?* (cross out 4 dolls)
- *Let's cross out the 4 dolls together.* Model crossing out the dolls as you count. Say: *Let's count the dolls as we cross them out: 1-2-3- ?* . (4) *We subtracted 4 dolls. Now let's count to see how many are left: 1-2-3-4-5- ?* (6) *How many dolls are left?* (6) *What does 10 − 4 equal?* (6)
- Point out that exercises 3–6 do not have pictures. Discuss strategies students can use to solve the problems, such as counting back, using related facts, modeling, visualizing, or drawing pictures and crossing out.
- Remind students that some problems, such as exercises 5 and 6, are written in columns instead of rows.
- Students who need more support subtracting may benefit from having manipulatives or number lines available.
- Have students complete exercises 2–6. Review students' work with them.

Power Practice

- Read the directions with students and make sure they understand what to do.
- Point out that exercises 11–20 do not have pictures. Discuss strategies students can use to subtract if they don't remember a fact. They can use related facts, count back, draw pictures, and cross out, model, and visualize.
- Have students complete the exercises. Then select a few exercises and have volunteers demonstrate subtracting.

Grade 1, Chapter 8, Cluster B **91**

CHALLENGE

Lesson Goal
- Identify coin amounts to 10¢ using pennies and nickels.

Introducing the Challenge
- Show students a penny and a nickel. Have students name each coin and tell you how much each is worth. Write 1¢ and 5¢ on the chalkboard. Have students read the amounts.
- Ask students what pennies and nickels are used for. Discuss the kinds of things students would like to buy with pennies and nickels.
- Explain to students that they are going to play a game. They will each have some pennies and nickels to buy things in a toy store.

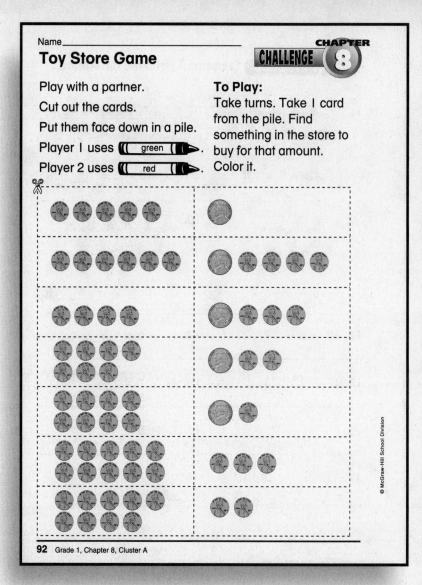

92 Grade 1, Chapter 8, Cluster A

Grade 1, Chapter 8, Cluster A 93

CHALLENGE

Using the Challenge

- Have students work with partners. They will need scissors, a red crayon, and a green crayon. (If you prefer, have students copy the coin cards onto index cards or onto strips of paper.)
- Tell students to cut out the coin cards, mix them up, and put them face down in a pile. Have each student take one of the crayons.
- Give each pair a game board. Tell students that they will take turns but that they need to decide who will go first.
- Explain that each player, in turn, takes a card from the pile. The player counts how much money he or she has. Then the player looks for something in the toy store that costs that much money. The player buys the item by coloring it with his or her crayon.
- When all the cards have been used once, they can be mixed up again and put face down in a pile. Tell students that if they get a card and there is nothing to buy for that amount of money, then the other player gets his or her turn. Have students play until there is nothing left in the store to buy.

Grade 1, Chapter 8, Cluster A 93

CHALLENGE

Lesson Goal
- Add with sums to 9.

Introducing the Challenge
- Ask students if they have ever done a word search puzzle. Explain that in a word search puzzle, words are hidden in rows and columns with other letters. Students draw a circle around the words they find.
- Show students the puzzle. Explain that this puzzle is a little different. It is an addition search. Instead of looking for words, they will look for addition facts.

Using the Challenge
- Read the directions with students.
- Work through the first example with students.
- Point out the first three numbers in the top row. Say: *Look at these numbers: 6, 3, 9. I know an addition fact that uses these numbers in this order: 6 + 3 = 9.* Write the addition fact on the chalkboard. Say: *I will put a circle around those numbers because they tell me an addition fact. Draw a circle around this fact.* Check students' work.
- *Now I will look across this row. Are there other facts I can find?* If necessary, help students locate the other facts in the row: 2 + 1 = 3 and 5 + 3 = 8. Have students circle the facts. Check their work.
- Say: *Now I want you to find the rest of the addition facts that are hidden in this puzzle. Look across one row at a time.*
- *Once you have finished all the rows, then look down the columns.* Point to the first column. Say: *There are some facts hidden in the columns too. Try to find as many addition facts as you can.* (There are 38 facts in all.)
- Some students who need more support may benefit from writing the + and = signs on the page to check their facts.

94 Grade 1, Chapter 8, Cluster B

Name_____

Addition Search

Circle the addition facts in the puzzle.

Look → and ↓.

6	3	9	4	3	2	8	7	2	1	3	4	5	3	8
5	4	9	6	4	4	9	1	2	1	8	9	6	1	2
2	5	7	8	7	3	1	8	5	3	4	2	6	7	7
1	2	3	9	4	4	8	5	7	1	1	2	5	8	9
8	1	9	3	6	5	9	2	6	8	9	6	2	8	7
6	3	4	6	8	4	1	5	5	5	2	7	8	4	3
3	2	7	5	1	4	5	1	3	4	3	4	3	6	9
7	5	5	1	7	9	4	6	1	7	1	5	6	5	5
3	3	8	6	9	2	6	3	5	7	9	4	5	9	3
1	9	2	9	1	1	2	5	6	8	2	8	5	3	9
4	5	2	5	3	7	3	8	4	3	7	1	2	3	5
3	5	2	6	5	4	5	2	4	6	3	6	4	7	8
2	8	4	7	2	9	7	2	1	4	3	7	3	3	6

© McGraw-Hill School Division

94 Grade 1, Chapter 8, Cluster B

Name_____

On Sale at the Toy Store

CHALLENGE CHAPTER 8

Each 😊 has [coins] for toys.

Cut out a toy for each to buy.

Paste it in the box.

Then write the subtraction problem to show how much each 😊 has left.

5¢

1. Tom has 10¢. He buys _____.

   10¢ – ____ = ____

7¢

2. Jan has 8¢. She buys _____.

   8¢ – ____ = ____

4¢

3. Mike has 12¢. He buys _____.

   12¢ – ____ = ____

3¢

4. Sue has 11¢. She buys _____.

   11¢ – ____ = ____

6¢

5. Kim has 9¢. She buys _____.

   9¢ – ____ = ____

8¢

CHALLENGE

Lesson Goal
- Subtract from 12.

Introducing the Challenge
- Ask students if they have ever had some money and paid for something they wanted themselves. Discuss what they bought. Ask: *What happened when you gave the store clerk your money? Did you get any change back?*
- Explain to students that on this next page some students have some money. Each of them can buy something on sale at the toy store. Students will decide which toy to buy and how much change, if any, the child will get back.

Using the Challenge
- Read the directions with students.
- Students will need pencils, scissors, and paste.
- Read the first example with students. Go over the choices of things that Tom can buy with his 10¢.
- Say: *You decide what Tom should buy with his money. For example, you may think Tom should buy the baseball. Then you would cut out the baseball and paste it in the sentence.*
- *Next, you would look at the price of the baseball and write the subtraction sentence to find out how much change Tom should get back. What subtraction sentence would you write if Tom bought the baseball?* (10 – 6) *What if he bought the dinosaur?* (10 – 4) *What if he bought the pencil?* (10 – 3)
- *Once you write the subtraction, you subtract to find his change. Write the difference on the line.*
- Have students begin the page. Some students who need more support may benefit from having manipulatives available.

Name_____

CHAPTER 9 — What Do I Need To Know?

Count to 20

Write how many.

1.

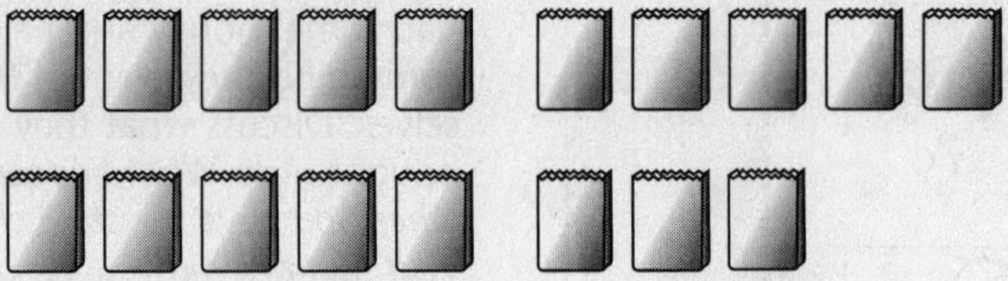

   _____

Patterns

Write the number that could come next in the pattern.

2. 4 4 8 4 4 8 4 4 8 4 4 ____

Add Sums to 12

Add. Write each sum.

3. 6 + 6 = _____

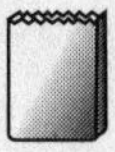

4. 4 + 5 = _____

5. 8
   + 2
   ___

95A Use with Grade 1, Chapter 9, Cluster A

Name_____

Subtract from 12

Subtract. Write each difference.

6. 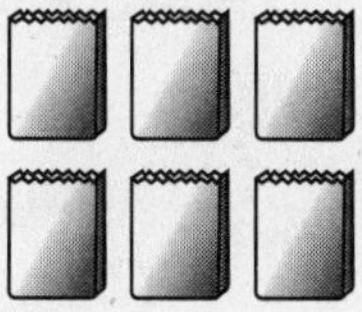 [picture of 6 bags]

    7. 9
       −1

    6 − 3 = _____

Related Facts

Complete two additions for each picture.

8. 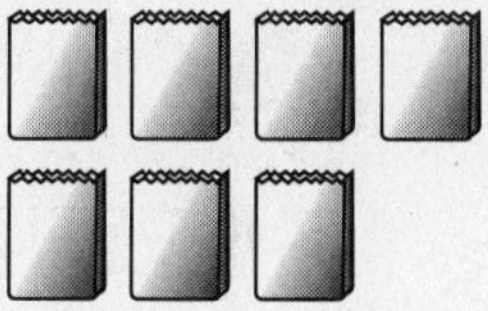 [picture of 6 bags]

    4 + 3 = _____

    3 + 4 = _____

Complete two subtractions for each picture.

9.  [picture of 8 bags]

    8 − 5 = _____

    8 − 3 = _____

Missing Addends

Write the missing number.

10. 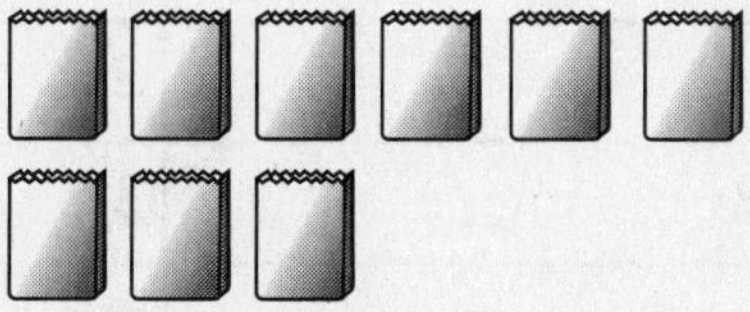 [picture of 8 bags]

    6 + ___ = 9

Use with Grade 1, Chapter 9, Cluster B   **95B**

CHAPTER 9 PRE-CHAPTER ASSESSMENT

Assessment Goal

This two-page assessment covers skills identified as necessary for success in Chapter 9 Addition and Subtraction Strategies and Facts to 20. The first page assesses the major prerequisite skills for Cluster A. The second page assesses the major prerequisite skills for Cluster B. When the Cluster A and Cluster B prerequisite skills overlap, the skill(s) will be covered in only one section.

Getting Started

- Allow students time to look over the two pages of the assessment. Point out the labels that identify the skills covered.
- Have students find math vocabulary terms used in the assessment. List vocabulary terms on the board as students identify them. If necessary, review the meanings of all essential math vocabulary.

Introducing the Assessment

- Explain to students that these pages will help you know if they are ready to start a new chapter in their math textbooks.
- Students who have transferred from another school may not have been introduced to some of these skills. Encourage students to do their best and assure them you will help them learn any needed skills.

Cluster A Challenge

Those students who demonstrate mastery of the skills on this page will not need to use the reteaching worksheets. Instead, these students can do the Cluster A Challenge found on pages 112-113.

Name_____

CHAPTER 9 — What Do I Need To Know?

Count to 20

Write how many.

1. [squares] __18__

Patterns

Write the number that could come next in the pattern.

2. 4 4 8 4 4 8 4 4 8 4 4 __8__

Add Sums to 12

Add. Write each sum.

3. [squares]   [squares]
   6 + 6 = __12__

4. $4 + 5 =$ __9__

5. 8
   $+2$
   $\overline{10}$

95A Use with Grade 1, Chapter 9, Cluster A

CLUSTER A PREREQUISITE SKILLS

The skills listed in this chart are those identified as major prerequisite skills for students' success in the lessons in Cluster A of the chapter. Each skill is covered by one or more assessment items as shown in the middle column. The right column provides the page numbers for the lessons in this book that reteach the Cluster A prerequisite skills.

Skill Name	Assessment Items	Lesson Pages
Count to 20	1	96
Patterns	2	97
Add Sums to 12	3-5	98-101

95C Grade 1, Chapter 9, Cluster A

Name_____

Subtract from 12

Subtract. Write each difference.

6. ▢▢▢
   ▢▢▢
   $6 - 3 =$ __3__

7. 9
   -1
   $\overline{8}$

Related Facts

Complete two additions for each picture.

8. ▢▢▢▢
   ▢▢▢
   $4 + 3 =$ __7__
   $3 + 4 =$ __7__

Complete two subtractions for each picture.

9. ▢▢▢▢▢
   ▢▢▢
   $8 - 5 =$ __3__
   $8 - 3 =$ __5__

Missing Addends

Write the missing number.

10. ▢▢▢▢▢▢
    ▢▢▢
    $6 +$ __3__ $= 9$

Use with Grade 1, Chapter 9, Cluster B **95B**

CHAPTER 9 PRE-CHAPTER ASSESSMENT

Alternative Assessment Strategies

- Oral administration of the assessment is appropriate for younger students or those whose native language is not English. Read the skills title and directions one section at a time. Check students' understanding by asking them to tell you how they will do the first exercise in the group.

- For some skill types you may wish to use group administration. In this technique, a small group or pair of students complete the assessment together. Through their discussion, you will be able to decide if supplementary reteaching materials are needed.

Intervention Materials

If students are not successful with the prerequisite skills assessed on these pages, reteaching lessons have been created to help them make the transition into the chapter.

Item correlation charts showing the skills lessons suitable for reteaching the prerequisite skills are found beneath the reproductions of each page of the assessment.

CLUSTER B PREREQUISITE SKILLS

The skills listed in this chart are those identified as major prerequisite skills for students' success in the lessons in Cluster B of the chapter. Each skill is covered by one or more assessment items as shown in the middle column. The right column provides the page numbers for the lessons in this book that reteach the Cluster B prerequisite skills

Skill Name	Assessment Items	Lesson Pages
Subtract from 12	6-7	102-105
Related Addition Facts	8	106-107
Related Subtraction Facts	9	108-109
Missing Addends	10	110-111

Cluster B Challenge

Those students who demonstrate mastery of the skills on this page will not need to use the reteaching worksheets. Instead, these students can do the Cluster B Challenge found on pages 114-115.

Grade 1, Chapter 9, Cluster B **95D**

USING THE LESSON

Lesson Goal
- Count objects and write the number in the group through 20.

What the Student Needs to Know
- Count objects through 10.
- Write numbers through 10.
- Count aloud through 20.

Getting Started
- Have the class count aloud to 20. Write the numbers on the chalkboard as students count.
- Point at random to the numbers 11 through 19. Have students say the names of these numbers.

What Can I Do?
Read the question and the response. Then read and discuss the example. Ask:
- *What are we counting in the example?* (shopping carts) *Why is there a number on each cart?* (To help us count the carts) Point out that the numbers go from left to right.
- Have students count the carts together, touching each cart as they say its number. Point out the writing line with the dotted answer. Have students trace over the 20. Ask: *What does this number show?* (the total number of carts in the group)

Try It
Read the directions aloud. Say:
- *Show me how you will count the carts for Exercise 1.* (Have a volunteer count the carts aloud.)
- *Where do you write the number?* (Make sure all students know where the writing lines are.)
- Have students complete exercise 2. Review their work.

Power Practice
- Read the directions aloud. Have students point to the writing lines.
- Have students complete the practice items. Review their answers.

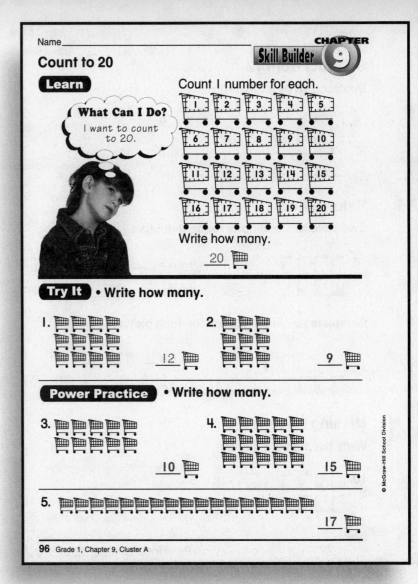

WHAT IF THE STUDENT CAN'T

Count Objects Through 10
- Give 10 counters to each pair of students. One student separates out a number of counters. The partner records the group on a sheet of paper using X's or circles for each counter. The students agree on the count. The second student writes the number. Partners switch roles.

Write Numbers Through 10
- Give students numbers made from felt or sandpaper. Have them trace the numbers with their fingers.
- Provide number cards with direction arrows showing how to write the numbers.

Count Aloud Through 20
- Write the number words for eleven to twenty on the board. Say each word as you point to it.
- Ask the student to write the corresponding number next to each word.
- Then have the student count aloud from 1 to 20.

Complete the Power Practice
- Have the student use counters to model each incorrect exercise. Ask the student to write the correct number on the line.

Name _____

CHAPTER 9 Skill Builder

Patterns

Learn

What Can I Do?
I want to find the next number in a pattern.

2 2 4 2 2 4 2 2 4 2 2 ___

Look for chunks.

|2 2 4|2 2 4|2 2 4|2 2 ___|

Find the pattern.

|2 2 4|2 2 4|2 2 4|2 2 ___|
↑
2 2 4 repeats.

Write the number that could come next in the pattern.

|2 2 4|2 2 4|2 2 4|2 2 4|

Try It • Write the number that could come next in the pattern.

1. |3 4 7|3 4 7|3 4 7|3 4 7|

2. 3 3 6 3 3 6 3 3 6 3 3 _6_

Power Practice • Write the number that could come next in the pattern.

3. 2 3 5 2 3 5 2 3 5 2 3 _5_

4. 4 4 8 4 4 8 4 4 8 4 4 _8_

5. 4 5 9 4 5 9 4 5 9 4 5 _9_

Grade 1, Chapter 9, Cluster A **97**

© McGraw-Hill School Division

WHAT IF THE STUDENT CAN'T

Read Numbers Through 9
- Write the numbers 0 through 9 on the chalkboard. Have students say each number aloud with you. Then write numbers at random for students to read.
- Students work in pairs using a set of 0–9 cards. One student chooses a card; the partner reads the number and tries to find that number of things in the classroom.

Identify the Repeating Unit in a Pattern
- Write a pattern based on the repeating unit A, B, C. Have students come up and add three-letter "chunks" to the pattern. Repeat with other patterns based on three letters.

Write Numbers Through 9
- Have students practice "air writing" to trace each of the numbers. Have students trace dotted-line numbers with writing arrows.

Complete the Power Practice
- For each incorrect exercise, have the student circle the repeating chunks in the pattern.
- Have the student identify the number that comes next, and write it on the line.

USING THE LESSON

Lesson Goal
- Write the next number in a repeating pattern.

What the Student Needs to Know
- Read numbers through 9.
- Identify the repeating unit in a pattern.
- Write numbers through 9.

Getting Started
- Have students clap and snap their fingers to join you in this pattern: clap, clap, snap; clap, clap, snap; and so on. That is, clap twice and snap your fingers, then repeat. Ask:
- *What pattern are we making?* (Clap twice and then snap.) *What three noises make up our pattern?* (clap, clap, snap)
- Draw the following pattern on the chalkboard: star, star, square; star, star, square. Have students continue the pattern across the chalkboard. Point out that this is like the clapping and snapping pattern: Three things repeat over and over.

What Can I Do?
Read the question and the response. Then read and discuss the example. Read the pattern with a pause after each chunk: 2, 2, 4 [pause] 2, 2, 4 [pause] 2, 2, 4 [pause]. Ask:
- *How many numbers are repeating?* (3 numbers) *What are the numbers?* (2, 2, 4)

Try It
Read the directions aloud. Read the first pattern with a pause after each chunk. Ask:
- *What chunk is repeating?* (347) *What number will you write on the writing line?* (7)
- Have students complete exercise 2. Check their work.

Power Practice
- Have students complete the practice items. Then review each answer.

Grade 1, Chapter 9, Cluster A **97**

USING THE LESSON

Lesson Goal
- Find sums to 12 by joining and counting groups.

What the Student Needs to Know
- Count objects to 9.
- Record addition using the plus sign.

Getting Started
- Place 5 red counters and 3 yellow counters on a table. Ask: *How many counters are there in each group?* (5 red; 3 yellow) *How can I add them?* (Put the groups together.) *How do you find the number in all?* (Count the objects in both groups.)
- *What sign do we use to show adding?* (the plus sign) Have a student demonstrate how to make the plus sign.
- Write the math sentence 5 + 3 = 8 to show the addition of the red and yellow counters.

What Can I Do?
Read the question and the response. Then read and discuss the example. Say:
- There are two boxes of tomatoes. How many tomatoes are there in each box? (6 and 4)
- What does the number sentence under the tomatoes show? (The two numbers are being added.)
- Let's count how many tomatoes there are in all: 1-2-3-4-5-6-7-8-9-10.
- Have students trace the number 10.

Try It
Read the directions aloud. Do the first exercise with students.
- Have students do the second exercise on their own. Point out that the problem is written in a column.

Power Practice
- Have students complete the practice items. Review each answer.

WHAT IF THE STUDENT CAN'T

Count Objects to 9
- Write the numbers from 1 to 9 on the chalkboard. Say each number as you point to it. Repeat, having the student count along with you.
- Have students work in pairs using fewer than 10 objects. One student displays a group of objects. The partner counts the objects and says the total. Students switch roles.

Record Addition Using the Plus Sign
- Model for students how to write the plus sign. Have them write the sign in the air and then write it on a piece of paper.
- Draw on the chalkboard: 3 objects with the number 3 beneath the group; 6 objects with the number 6 beneath the group. Ask the student to write a plus sign between the numbers and then write the sum.

Complete the Power Practice
- Provide counters so that students can model any incorrect exercises.
- Have students write the correct sum for each.

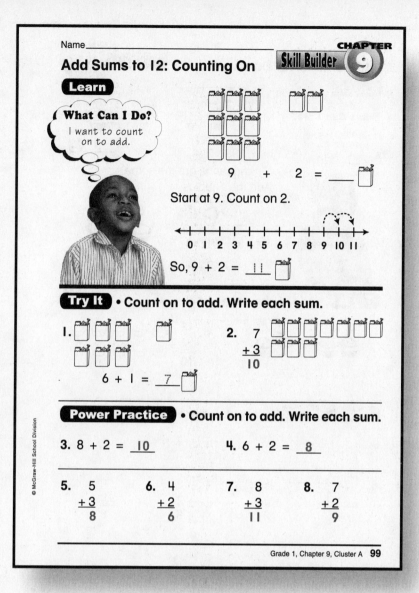

Using the Lesson

Lesson Goal
- Find sums to 12 by counting on.

What the Student Needs to Know
- Count aloud to 12.
- Count objects to 9.
- Read the plus and equal signs.

Getting Started
- On the chalkboard, draw a 0–10 number line. Display a group of 7 red counters and a group of 2 yellow counters. Ask:
- *How many counters are in the first group?* (7) *Let's find 7 on the number line.*
- *How many counters are in the second group?* (2) *Let's use the number line to find how many counters there are in all.*
- *Let's count on 2 from 7. What number are we at?* (9)
- So, 7 + 2 = 9.

What Can I Do?
Read the question and the response. Then read and discuss the example. Ask:
- *How many bags are there in each group?* (9 in the first group; 2 in the second group) *How many bags are there in all?* (11) Have students count all the bags to confirm that there are 11 in all and write 11 on the line.
- *How can we find the sum without saying all the numbers up to eleven?* (Start the counting at 9; then count on 2)

Try It
Read the directions aloud. Do the first exercise with the students counting on from 6. Have students trace the answer.
- Have students complete exercise 2 and check their answers.

Power Practice
- Have students complete the practice items. Tell students to count on silently to find the sums. Then review their answers.

Grade 1, Chapter 9, Cluster A **99**

What If The Student Can't

Count Aloud to 12
- Write the numbers 0-12 on the chalk board, saying each number as you write it. Then have the student come up and point to each number as he or she counts to 12.

Count Objects to 9
- Use dot cards showing the numbers 1 through 9. Mix up the cards. The student picks a card and counts the dots aloud.
- Provide number cards for 0 through 9. Have students work in pairs. One student chooses a card; the partner draws a picture of that number of objects.

Read the Plus and Equal Signs
- Write the word *plus* and the symbol + on the chalkboard. Repeat for the word *equals* and the = sign. Have students write a few addition sentences as you dictate them.
- Give pairs of students one set of 0–9 cards and cards with the plus sign and equal sign. Have students work together to make true sentences.

Complete the Power Practice
- Have the student use a number line to count on for each incorrect exercise. Then have the student write the correct answer for each one.

USING THE LESSON

Lesson Goal
- Find doubles to 12.

What the Student Needs to Know
- Recognize addition as joining two groups.
- Use the plus and equal signs.
- Use counters to model an addition problem.

Getting Started
- Draw 4 stars on the chalkboard. Ask: *How many stars do you see?* (4) *Who can draw another group with the same number?* Have a student come to the chalkboard and draw 4 more stars.
- *If you add two numbers that are the same, we say you are adding doubles. What doubles are we adding?* (4 + 4) *What is the sum?* (8)
- *What are some other doubles you know?* (Answers will vary. Write students' examples on the chalkboard.)

What Can I Do?
Read the question and the response. Then read and discuss the example. Point out the writing line for the first sum. Ask:
- *What sum should go on the line?* (12) If any students look uncertain, have them count the muffins to confirm the sum.
- *How do you know that this problem is a double?* (Both numbers being added are the same.)

Try It
- Read the directions aloud and do the first exercise with students. Have them add the doubles and trace the answer.
- Have them complete exercise 2. Check their answer.

Power Practice
- Have students complete the practice items. Then review each answer.

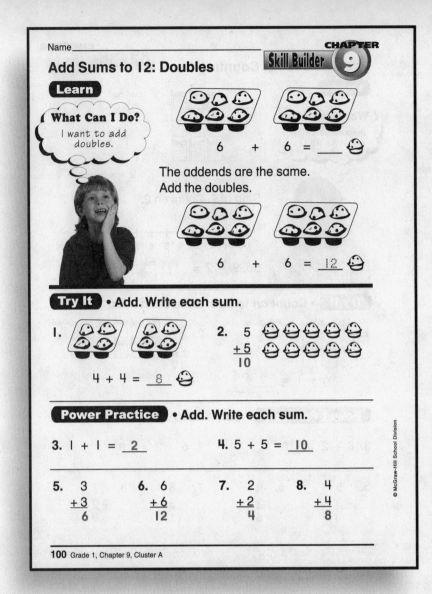

WHAT IF THE STUDENT CAN'T

Recognize Addition as Joining Two Groups
- Write 3 + 4 = 7 on the board. Ask: *Can you make up a story that uses these numbers?* As the student tells a story, draw groups on the chalkboard to go along with it. Point out that all addition stories are the same because groups of things are being put together.

Use the Plus and Equals Signs
- Say the addition sentence: 4 + 1 = 5. Ask the student to write the sentence. Repeat with other examples.

Use Counters to Model an Addition Problem
- Use number cards to 9. Put the cards for 5 and 3 on a table. Have students count out two groups of objects and put them under the cards. Put a card with the plus sign between the two numbers. Ask the student to push the objects together, count them, and tell the sum. Repeat with other examples.

Complete the Power Practice
- Provide counters so that students can model any incorrect exercises. Have students write the correct answer for each.

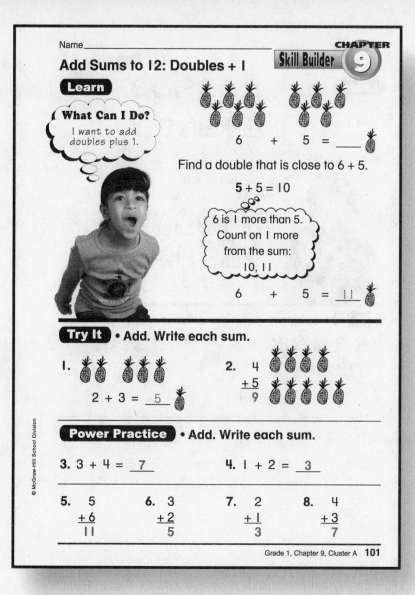

Using the Lesson

Lesson Goal
- Find sums to 12 by using the strategy doubles +1.

What the Student Needs to Know
- Add doubles through 12.
- Count on by 1.
- Identify two numbers that are 1 apart.

Getting Started
- Give pairs of students 3 red counters and 3 yellow counters. Remind students that adding two same numbers is called doubles. Ask: *What is the sum of this double?* (6) Give each pair one more red counter. *Now what is the sum?* (7)
- Demonstrate the reasoning for students: 3 + 3 = 6, so 4 + 3 = 7. Explain that if you add 1 to one of the numbers, you also add 1 to the sum. Repeat with one or two more examples.

What Can I Do?
Read the question and the response. Then read and discuss the example. Ask:
- *What addition problem is shown?* (6 + 5) *What double is close to 6 + 5, but less?* (5 + 5) *How can you use the double to find 6 + 5?* (5 + 5 is 10, so 6 + 5 is 11)

Try It
Read the directions aloud and do the first exercise with the students. Ask: *What double will you use to help solve this problem?* (2 + 2) Say: that's right; if 2 + 2 = 4, 2 + 3 = 5.
- Have students complete exercise 2. Check their answers.

Power Practice
- Have students tell the double that will help them solve each exercise and then complete the practice items. Review the answers.

USING THE LESSON

Lesson Goal
- Subtract from numbers through 12.

What the Student Needs to Know
- Count objects to 9.
- Recognize subtraction as taking away objects.
- Record subtraction using the minus sign.

Getting Started
- Provide students with 9 counters and a sheet of paper. Have each student count to verify the total and write 9 on their papers. Then say:
- *Push 3 objects to one side. How many objects are left?* (6) *You have just subtracted 3 from 9. What subtraction fact can you write?* Students write the fact in either horizontal or vertical form:
9 – 3 = 6

$$\begin{array}{r}9\\-3\\\hline 6\end{array}$$

- Have students use workmats divided in half with a horizontal line. Following your directions, they put 8 objects above the line. Then they subtract 3 objects by moving them below the line.

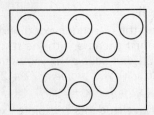

Ask students to write the subtraction fact.

8 – 3 = 5

Repeat with other subtraction facts.

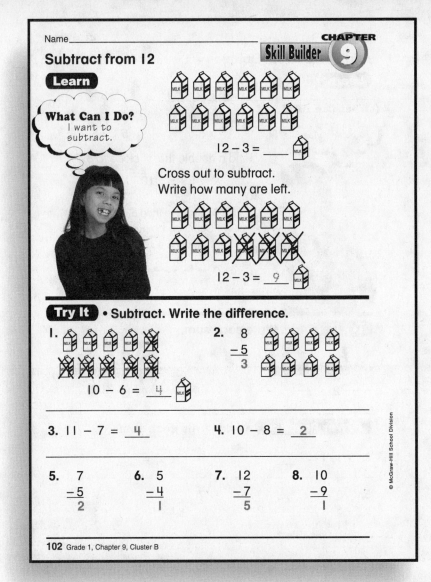

WHAT IF THE STUDENT CAN'T

Count Objects to 9
- Write the numbers 1 through 9 in a row on the chalkboard. For each number, ask students to draw a dot picture. Then have the student count the dots in each picture.
- Students work in pairs using objects. One student puts out a number of objects less than 10; the partner counts the objects and says the total.

Recognize Subtraction as Taking away
- Give the student 4 red counters and 7 blue counters. Have students count the total number. Then ask the student to separate out the red counters and put them aside. Ask the student how many counters have been subtracted from 11. Ask how many counters are left.
- Draw a large fence on the chalkboard. Have several students take turns drawing birds on the fence until there are 8 birds. Say: *3 birds flew away. Who can change the drawing to show this?* (A student erases 3 birds.) Repeat with other examples.

102 Grade 1, Chapter 9, Cluster B

Name_____

Power Practice • Subtract. Write the difference.

9. 8 − 6 = __2__

10. 10 − 4 = __6__

11. 12
    − 5
    ——
     7

12. 11
    − 5
    ——
     6

13. 11 − 9 = __2__

14. 10 − 7 = __3__

15. 8 − 7 = __1__

16. 12 − 8 = __4__

17. 7 − 4 = __3__

18. 11 − 6 = __5__

19. 11
    − 4
    ——
     7

20. 7
    − 6
    ——
     1

21. 12
    − 9
    ——
     3

22. 6
    − 4
    ——
     2

23. 12
    − 4
    ——
     8

24. 6
    − 5
    ——
     1

25. 5
    − 4
    ——
     1

26. 11
    − 8
    ——
     3

Grade 1, Chapter 9, Cluster B **103**

WHAT IF THE STUDENT CAN'T

Record Subtraction Using the Minus Sign

- Write on the chalkboard: 6 − 1 = 5. Have students read the sentence in as many different ways as they can: six minus one equals five, six take away one is five, one subtracted from six is five, the difference between six and one is five.
- Dictate subtraction problems to the student, using a variety of styles as stated in the activity above. For each dictation, students write the subtraction sentence.

Complete the Power Practice

- Have the student use counters or draw pictures and cross out to model each problem that was done incorrectly.
- Have the student write the correct answer for each one.

USING THE LESSON

What Can I Do?

- Have students read the question and the response. Then read and discuss the example. Ask:
- *How many cartons of milk are in the top group?* (12; if necessary, have students count to confirm this.) *How many cartons do we want subtract?* (3)
- Have students cross out 3 cartons and count the number left. Have students trace the 9 in the answer.

Try It

- Read the directions aloud. Do the first exercise with the students. Have students trace over the 6 dotted-line Xs, count the number left, and write it on the line.
- Have students complete exercises 2–8. Check their work.

Power Practice

- Have students complete the practice items.
- When students have finished, ask volunteers to come up to the chalkboard and draw pictures for some of exercises 13–26, and show how they solved them.

Learn with Partners & Parents

- Students draw a large picture of animals around a pond. They can include frogs, birds, ducks or squirrels. There should be 6 to 12 of each kind of animal.
- Students then tell subtraction stories using their pictures. For example, 8 squirrels were eating nuts in a tree by the pond. Two people came walking by. Three squirrels got frightened and ran away so there were 5 squirrels left.

Grade 1, Chapter 9, Cluster B **103**

USING THE LESSON

Lesson Goal
- Subtract from numbers through 12 by counting back.

What the Student Needs to Know
- Count backwards aloud from 12.
- Count objects to 9.
- Read the minus and equal signs.

Getting Started
- Have the class start at 10 and count back to 0. Draw a 0–10 number line on the chalkboard. Point to various numbers and have students count back from those starting points.
- Point to 9 on the number line. Ask students to count back. Point out that they have just subtracted 2 from 9. Ask: *What is 9 – 2?* (7)

What Can I Do?
- Read the question and the response. Then read and discuss the example. Ask:
- *What number goes on the writing line?* (7) *What are two ways to find the answer?* (Cross out 3 objects in the top group; count back 3 on the number line.)

Try It
- Read the directions aloud. Do the first exercise with students, using the number line.
- Have Students complete exercise 2. Check their answers.

Power Practice
- Have students complete the practice items. Then review each answer. Ask students to explain how they found the differences.

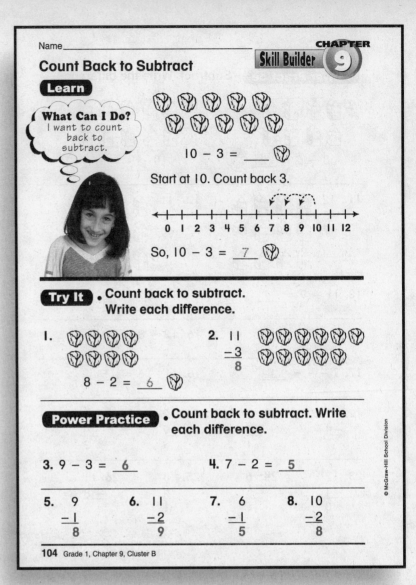

WHAT IF THE STUDENT CAN'T

Count Backwards From 12
- Write the numbers 0 through 12 on the chalkboard, saying each number as you write it. Have the student count back, starting with 12.
- Have the student think of a situation in which someone counts backwards. Have the student draw a picture of it and write the numbers on the picture.

Count Objects to 9
- Give pairs of students dot cards for 1–9. Each student picks a card, counts the dots, and tells the number. The student with the greater number keeps both cards. Repeat until the cards are gone.

Read the Minus and Equal Signs
- Write the word minus and the symbol "–" on the board. Repeat for the word equals and the = sign. Have students use the symbols to write a few subtraction facts.

Complete the Power Practice
- Have the student use a number line to rework each incorrect exercise.
- Have the student write the correct answer.

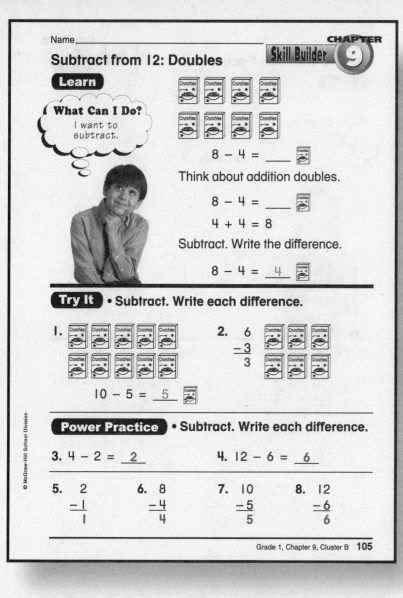

USING THE LESSON

Lesson Goal
- Subtract from numbers through 12 using doubles.

What the Student Needs to Know
- Add doubles with sums through 12.
- Use the minus and equal signs.
- Use counters to model a subtraction problem.

Getting Started
- Ask six students to come up to the chalk board. Give each student a number from 1–6. The student draws that many circles in a row. Say: *Now draw another row underneath. Draw the same number of circles in the second row.*
- Point out that the drawings show addition doubles. Have students say the sums for the models.
- Ask the students at the chalkboard to cross out the circles in the second row. Ask: *What subtraction problems have we shown?* (12 – 6 = 6, 10 – 5 = 5, 8 – 4 = 4, 6 – 3 = 3, 4 – 2 = 2, 2 – 1 = 1)

What Can I Do?
Read the question and the response. Then discuss the example. Ask:
- *What double does the picture show?* (4 + 4 = 8) *If you take away the second row, how many cereal boxes are left?* (4) Have students cross out the boxes in the bottom row.
- Say: *If 4 + 4 = 8, then 8 – 4 = 4. Trace the answer, 4.*

Try It
- Do the first exercise with students.
- Have students complete exercise 2. Check their work.

Power Practice
- Have students complete the practice items. Then review each answer.

WHAT IF THE STUDENT CAN'T

Addition Doubles Through 12
- Students work in pairs using counters of two colors. One student puts out a row of up to 6 counters; the partner uses the other color to match the row. Students say and then write the addition double for the model.

Use the Minus and Equal Signs
- Write this word sentence: "Six minus two equals four." Ask students to write the sentence another way.
6 – 2 = 4.
Repeat with other examples.

Use Counters to Model a Subtraction Problem
- Use number cards through 12. Place the cards for 7 and 4 on a table. Put a card with the minus sign between the two number cards. Have students count out 7 objects and then take 4 away. The student counts the number left and tells the difference. Repeat with some other examples.

Complete the Power Practice
- Provide counters so that students can model any incorrect exercises.
- Have students write the correct answers for each one.

USING THE LESSON

Lesson Goal
- Complete pairs of related addition facts.

What the Student Needs to Know
- Read and write numbers through 12.
- Use connecting cubes or counters to model addition facts.
- Read vertical and horizontal addition sentences.

Getting Started
- Have students work in small groups using connecting cubes or counters. Each student makes an addition fact using two colors. Ask:
- *Can you tell me two facts for your model?* (Yes, unless both numbers added are the same.)
- Have several volunteers write their related addition facts on the chalkboard. Then have students in each group take turns telling their related fact pairs.
- Write this fact pair on the chalkboard:

  4 + 3 = ?
  3 + 4 = ?

  Ask: *What is the same about the facts?* (The same two numbers are being added; the sum is the same.) *What is different?* (The order of the numbers are switched.)
- Stand facing the class. Hold up 2 fingers on the left hand and 4 fingers on the right hand. Have students call out the fact. (2 + 4 = 6) Turn around so your back is to the class. Have students call out the new fact. (4 + 2 = 6)

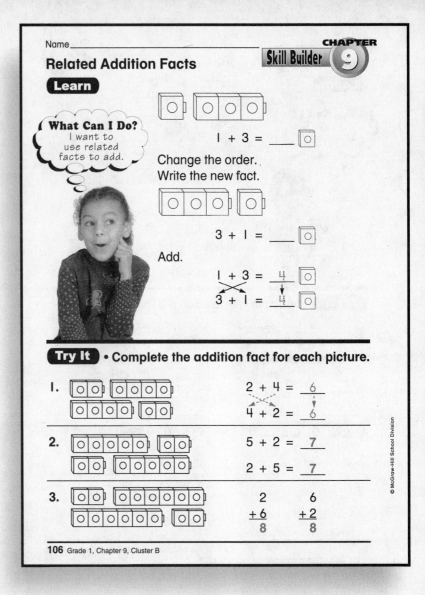

WHAT IF THE STUDENT CAN'T

Read and Write Numbers Through 12
- Draw a 0–12 number line on the chalk board. Have students use the number line to count aloud to 12.
- For several of the numbers from 0 through 12, have students find that number of objects in the classroom.
- Write the word names from *one* through *twelve* on the board. Have students write the corresponding number next to each word name.

Use Connecting Cubes or Counters to Model Addition Facts
- Have students work in pairs. One student shows a fact with counters or connecting cubes; the partner says the fact and then counts to find the sum. Have students switch roles.
- Have students work in pairs, using addition flash cards. Each student picks a card and then shows the fact with counters or connecting cubes. Have students check each other's work.

106 Grade 1, Chapter 9, Cluster B

Name _____

Power Practice • Complete the addition fact for each picture.

4. 4 + 6 = 10
   6 + 4 = 10

5. 7 + 1 = 8
   1 + 7 = 8

6. 4   7
   +7   +4
   11   11

7. 2   8
   +8   +2
   10   10

8. 9 + 3 = 12   9. 5 + 4 = 9
   3 + 9 = 12   4 + 5 = 9

10. 3 + 5 = 8   11. 7 + 2 = 9
    5 + 3 = 8   2 + 7 = 9

12. 8   1   13. 9   2   14. 4   8
    +1 +8   +2   +9   +8   +4
     9   9   11   11   12   12

Grade 1, Chapter 9, Cluster B 107

USING THE LESSON

What Can I Do?
Read the question and the response. Then read and discuss the example. Ask:

- *Which fact is shown by the first picture?* (1 + 3 = 4) *Which fact is shown by the second picture?* (3 + 1 = 4) *How are these facts alike?* (The same numbers, 1 and 3, are being added; the sum is the same.) *How are these facts different?* (The order of the numbers being added is different.)
- *What do pairs of facts like this prove?* (You can add two numbers in either order without changing the sum.)

Try It
Read the directions aloud. Have students look at the pictures and tell you the fact pairs for each exercise before they begin. Ask:

- *What will be true of the facts in each exercise?* (The sums will be the same.)
- After students have finished the exercises, check their answers.

Power Practice
- Have students complete the practice exercises. Then review students' answers.
- Summarize the exercise by asking: *What happens when you switch the order of the numbers when you are adding?* (The sum remains the same.)

WHAT IF THE STUDENT CAN'T

Read Vertical and Horizontal Addition Sentences
- Write the following addition sentence on the chalkboard: 6 + 1 = 7. Ask the students to read the sentence pointing to each number or symbols as he or she says it. Then ask the student to write the sentence in vertical form.
- Have students work in pairs. One student uses counters and shows an addition fact; the other student counts to find the sum; one student writes the fact as a horizontal number sentence; the other student writes the fact in vertical form.

Complete the Power Practice
- For each incorrect exercise, have the student use corrective cubes to model addition facts. Then have the students write the correct answers.

Grade 1, Chapter 9, Cluster B 107

USING THE LESSON

Lesson Goal
- Complete pairs of related subtraction facts.

What the Student Needs to Know
- Read and write numbers through 12.
- Use counters to model subtraction facts.
- Read vertical and horizontal subtraction sentences.

Getting Started
- Display 7 counters. Separate them into a group of 3 and a group of 4. Ask: *If you take away 4 counters, how many are left?* (3) *If you take away 3 counters, how many are left?* (4) *What two subtraction facts does this model show?* (7 − 4 = 3, 7 − 3 = 4)
- Using the example with 7 counters, introduce the terms *whole* and *part*. The whole group has 7, so the whole is 7. There are two parts, 3 and 4. Write the following on the chalkboard:

  7 − 4 = 3

  7 − 3 = 4

  Ask: *What is the same about these facts?* (They have the same whole and the same two parts.)

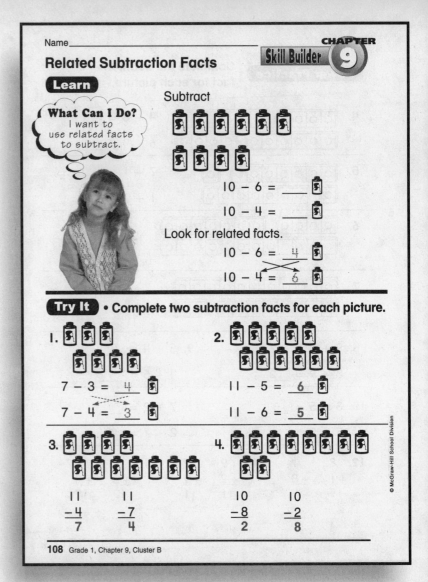

WHAT IF THE STUDENT CAN'T

Read and Write Numbers Through 12
- Have the student count aloud to 12. Draw a number line showing the numbers from 0 through 12. Point to various numbers and have the student say them.
- Write the word names from *one* through *twelve* in a vertical column on the chalkboard. Have the student write the correct number next to each word name.

Use Counters to Model Subtraction Facts
- Remind the student that subtraction means taking some away and finding the number left. Ask the student to show a subtraction fact with counters by displaying a group and then taking some away. Ask: *How many are left?* Then have the student write the subtraction fact.
- Have students work in pairs. One student shows a subtraction fact with counters; the partner then writes the fact.

108 Grade 1, Chapter 9, Cluster B

Name_____

Power Practice • Complete two subtraction facts for each picture.

5. $9 - 3 = \underline{6}$
    $9 - 6 = \underline{3}$

6. $6 - 4 = \underline{2}$
    $6 - 2 = \underline{4}$

7. $\begin{array}{r} 12 \\ -9 \\ \hline 3 \end{array}$ $\begin{array}{r} 12 \\ -3 \\ \hline 9 \end{array}$

8. $\begin{array}{r} 8 \\ -3 \\ \hline 5 \end{array}$ $\begin{array}{r} 8 \\ -5 \\ \hline 3 \end{array}$

9. $5 - 3 = \underline{2}$
    $5 - 2 = \underline{3}$

10. $10 - 9 = \underline{1}$
    $10 - 1 = \underline{9}$

11. $12 - 5 = \underline{7}$
    $12 - 7 = \underline{5}$

12. $9 - 5 = \underline{4}$
    $9 - 4 = \underline{5}$

13. $\begin{array}{r} 11 \\ -8 \\ \hline 3 \end{array}$ $\begin{array}{r} 11 \\ -3 \\ \hline 8 \end{array}$

14. $\begin{array}{r} 7 \\ -2 \\ \hline 5 \end{array}$ $\begin{array}{r} 7 \\ -5 \\ \hline 2 \end{array}$

15. $\begin{array}{r} 12 \\ -4 \\ \hline 8 \end{array}$ $\begin{array}{r} 12 \\ -8 \\ \hline 4 \end{array}$

Grade 1, Chapter 9, Cluster B **109**

WHAT IF THE STUDENT CAN'T

Read Vertical and Horizontal Subtraction Sentences

- Write the following on the chalkboard: 8 − 2 = 6. Ask the student to read the fact, pointing to each number or symbol as he or she says it. Ask the student to write its vertical form. Continue with other subtraction facts.

- Have students work in pairs. One student uses counters and takes some away to show a subtraction fact; the other student counts to find the difference; one student writes the fact in horizontal form with the equals sign; the other student writes the fact in vertical form.

Complete the Power Practice

- Review each incorrect answer. Have the student use counters or count back to rework the problems. Then have the student write the correct answers.

USING THE LESSON

What Can I Do?

- Read the question and the response. Then read and discuss the example. Ask:
  What two subtraction facts does this picture show? (10 − 6 and 10 − 4.)

- *What is the answer to the first fact, 10 − 6?* (4) *What is the answer to the second fact, 10 − 4?* (6) *The arrow shows the two facts are related.*

Try It

- Read the directions aloud and complete the exercise with students. Have students subtract the top row by putting their finger over it. Ask: *How many did you subtract?* (3) *How many are left?* (4) *What subtraction fact is this?* (7 − 3 = 4) Have students subtract the bottom row in the same way to get the second fact in the pair.

- Have students complete exercises 2–4. Check their answers.

Power Practice

- Read the directions aloud. Ask: *How can you use the pictures to help you do exercises 5–8?* (Students can cover up the top row to subtract. Then they can cover up the bottom row. This will give them the two related subtraction facts.)

- Have the students complete the practice items. As they work on the last three rows, remind them that the same three numbers must be used in each fact in a pair.

Grade 1, Chapter 9, Cluster B **109**

USING THE LESSON

Lesson Goal
- Use related subtraction facts to find missing addends.

What the Student Needs to Know
- Write a related subtraction fact for an addition fact.
- Use counters or connecting cubes to model addition and subtraction facts.
- Subtract numbers through 12.

Getting Started
- Write the term *addend* on the chalkboard. Cover up the suffix *-end* and ask: *What word do you see now?* (add) Explain that an addend is a number that we add. Write the addition fact 4 + 5 = 9 and have students identify the two addends.
- Then write the following:
  □ + 5 = 9

  Ask: *What is the missing addend?* (4)
- Explain to students that another way to find a missing addend is by subtracting. Write the following on the chalkboard:
  9 − 5 = 4
- Say: *You can find a missing addend by the subtracting the other addend from the sum.*

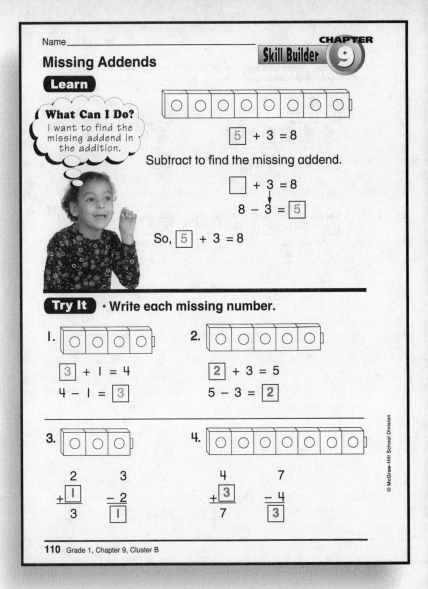

WHAT IF THE STUDENT CAN'T

Write a Related Subtraction Fact for an Addition Fact
- Have the student use part-part-whole boxes like this:

Have the students begin by using counters for the parts and the whole. Then move on to numbers. For each model, ask students to write two addition facts and two subtraction facts.

- Have students work in pairs using number cards. Each pair also needs cards for plus, minus, and equal signs. One student uses the cards to make an addition fact; the partner uses the cards to make a related subtraction fact.

Use Counters or Connecting Cubes to Model Addition and Subtraction Facts
- Review the meaning of addition as the joining of two groups. Have students work in pairs using counters. Each student displays a few counters. This is recorded as two numbers with a plus sign. Students push the counters together to find and write the sum.

110 Grade 1, Chapter 9, Cluster B

Name_____

Power Practice • Write each missing number.

5. $\boxed{4} + 1 = 5$
   $5 - 1 = \boxed{4}$

6. $\boxed{5} + 2 = 7$
   $7 - 2 = \boxed{5}$

7. $\begin{array}{r}1\\+\boxed{3}\\\hline 4\end{array}$ $\begin{array}{r}4\\-1\\\hline \boxed{3}\end{array}$

8. $\begin{array}{r}2\\+\boxed{4}\\\hline 6\end{array}$ $\begin{array}{r}6\\-2\\\hline \boxed{4}\end{array}$

9. $\boxed{6} + 2 = 8$
   $8 - 2 = \boxed{6}$

10. $\boxed{4} + 6 = 10$
    $10 - 6 = \boxed{4}$

11. $\boxed{6} + 5 = 11$
    $11 - 5 = \boxed{6}$

12. $\boxed{6} + 3 = 9$
    $9 - 3 = \boxed{6}$

13. $\begin{array}{r}6\\+\boxed{2}\\\hline 8\end{array}$ $\begin{array}{r}8\\-2\\\hline 6\end{array}$

14. $\begin{array}{r}3\\+\boxed{7}\\\hline 10\end{array}$ $\begin{array}{r}10\\-7\\\hline 3\end{array}$

15. $\begin{array}{r}4\\+\boxed{5}\\\hline 9\end{array}$ $\begin{array}{r}9\\-4\\\hline 5\end{array}$

Grade 1, Chapter 9, Cluster B **111**

WHAT IF THE STUDENT CAN'T

- Review the meaning of subtraction as taking part of a group away. Students work in pairs. One student displays some counters; the partner takes some away. Students find the number left and write the fact.

- Students work in pairs using two 1–6 number cubes. Each student rolls a cube. Then they use the two numbers to write a subtraction fact. Provide counters for students to use to confirm answers.

Subtract Numbers Through 12

- Have the student draw a large tree with 12 birds in it. Provide sticky notes. Have the student "take away" birds by covering them up with the notes.

Complete the Power Practice

- Have the student use connecting cubes or draw a picture to model each incorrect exercise. Then have the student write the correct answer for each exercise.

USING THE LESSON

What Can I Do?

- Read the question and the response. Then read and discuss the example.

  Say: *One addend is missing. The empty box shows the missing addend. How can you find the missing addend?* (Subtract 3 from 8.)

- Review the last part of the example, helping students to recognize logical thinking.

  Say: *To find the missing addend in $\square + 3 = 8$, think, "What number + 3 is 8? Since 8 − 3 = 5, it must be true that 5 + 3 = 8."*

Try It

- Read the directions aloud. Lead students through exercise 1. Have them read the first sentence as "What number plus 1 equals 4?" Direct students to color 1 cube on the right side of the cube train. Ask: *What two parts are shown now?* (3 and 1) *What addition fact does the picture show?* (3 + 1 = 4) Have students then complete the related subtraction fact, 4 − 1 = 3.

- Have students compete exercises 2–4. Check their work.

Power Practice

- Read the directions aloud. Have students complete the practice exercises. Review students' answers.

Grade 1, Chapter 9, Cluster B **111**

CHALLENGE

Lesson Goal
- Use addition facts through 12 to move spaces on a game board.

Introducing the Challenge
- Have students look at the game board on page 113. Relate the board to students' experiences at grocery stores:
- *Why do you get a shopping cart?* (to put all your groceries in it) *What happens to the cart as you go around the store?* (It gets filled with groceries.)
- Have students look at the foods on the game board. Explain that if they land on a space, they will have chosen that food.
- Now have students look at the game cards on page 112. These cards will tell them how many spaces they can move on each turn. Point out that the number of spaces to move is *not* the same as the sum for the addition fact.
- Have students cut out the game cards.
- Establish how to cross the checkout line on the game board to complete the game. It is easiest for players to "check out" if the number they have is the same as or greater than the one needed to finish. For example, if a player is one space away, he or she can finish with any of the move cards.

Name_____

Go Shopping!

Cut out the cards.
Mix them up.
Put them face down in a pile.
Play with a partner.

To play:
Take turns.
Turn over 1 card. Add.
If you are correct, move the number of spaces shown.

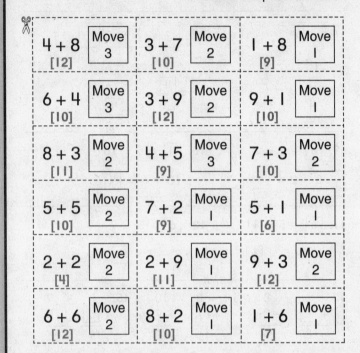

112 Grade 1, Chapter 9, Cluster A

Name_____

Grade 1, Chapter 9, Cluster A **113**

CHALLENGE

Using the Challenge

- Read the game directions at the top of page 112. Check that students understand how to play the game. Ask: *What must you do after you draw a card?* (Answer the addition problem correctly.) *If you get the answer right, then what do you do?* (Move the number of spaces shown on the card.)

- Each player will need a game marker. Counters or other small objects can be used.

- Have students run a finger along the game board path, noting the direction of movement. They start at the top where the empty shopping cart is shown. Then they wind around the path until they get to the checkout counter.

- Have the class agree what to do with cards that have been drawn. They can be returned to the pile, or students can hold them until one player reaches the checkout.

- If one student thinks the partner answered an addition fact incorrectly, the first player challenges the partner to prove the answer using counters. If this does not resolve the disagreement, the players ask you to check.

- After students are familiar with the game, suggest that they keep track of what foods they are "buying" during each round of the game.

Grade 1, Chapter 9, Cluster A **113**

CHALLENGE

Lesson Goal
- Use subtraction facts through 12 to move spaces on a game board.

Introducing the Challenge
- Have students first look at the game board on page 115. Relate the board to students' experiences at a shopping mall.
- *What are some of your favorite stores at the shopping mall? Which stores on the page would you like to visit?*
- *What do you see at the end of the path?* (a food court) Check that students know this is a place with restaurants and tables.
- Have students run a finger along the game board path, noting the direction of movement. They start at the top left where the word "Start" is printed. Then they wind around the path until they get to the food court.
- Now have students look at the spinner on page 114. Each student needs a pencil and large paper clip for the spinner. Demonstrate how to spin the paper clip around the point of the pencil. Explain that the spinner tells them how many spaces they can move on each turn.
- Establish what happens as students get near the food court on the game board. It is easiest if players can win by spinning a number equal to or greater than the one needed. For example, if a player is one space away, he or she can finish with any of the numbers on the spinner.

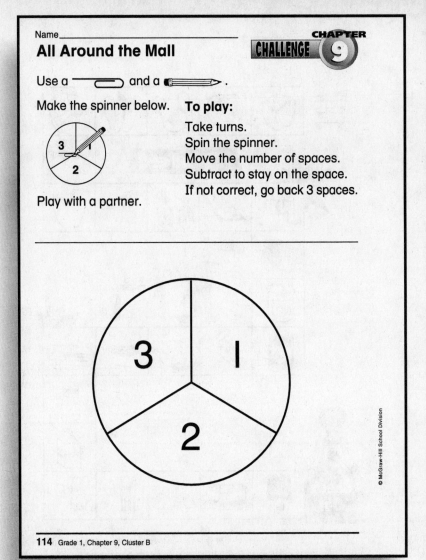

Name_____

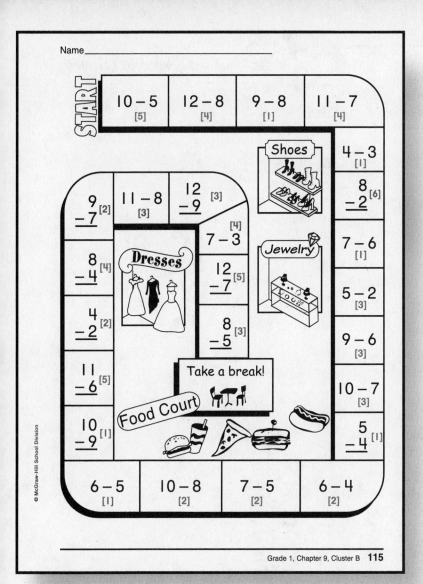

Grade 1, Chapter 9, Cluster B **115**

CHALLENGE

Using the Challenge

- Each player will need a game marker. Counters or other small objects can be used.
- Read the game directions at the top of page 114. Check that students understand how to play the game. Ask: *What do you do first when it is your turn?* (Spin the spinner.) *What do you do next?* (Move that number of spaces and say the answer to the subtraction fact.) *What happens if your answer to the subtraction fact is wrong?* (You must move back 3 spaces.)
- If one student thinks the partner has given an incorrect answer to a subtraction fact, the first player challenges the partner to prove the answer using counters. If this does not resolve the disagreement, the players ask you to check.
- Remind students that they use the spinner to find the number of spaces to move on their turn. They do not use the answers to the subtraction facts to determine the moves.
- After students are familiar with the game, suggest that they keep track of which stores they "visit" during each round of the game.

Grade 1, Chapter 9, Cluster B **115**

Name_____

CHAPTER 10 — What Do I Need To Know?

Count to 100

Write how many.

1.  _____

Same Number

Write how many.
Circle the same numbers.

2.

_____   _____   _____

Before and After

Write who comes before and after.

Sara   Emma   Mike   Wendy

3. Before Emma _____

4. After Mike _____

115A Use with Grade 1, Chapter 10, Cluster A

Name_____

Ordinals

Write the name of the person.

Fred Jan Mark Nia Olivia

5. I am first. _____

Order Numbers to 100

Write each number that comes next.

6. 9 10 11 ☐   **7.** 27 28 29 ☐

Estimate Time

Circle the time it would take to do each activity.

8. I brush my teeth.
about 1 minute
about 1 hour

9. I go to camp.
about 1 hour
about 1 day

10. I play a game.
about 1 minute
about 1 hour

CHAPTER 10
PRE-CHAPTER ASSESSMENT

Assessment Goal
This two-page assessment covers skills identified as necessary for success in Chapter 10 Time. The first page assesses the major prerequisite skills for Cluster A. The second page assesses the major prerequisite skills for Cluster B. When the Cluster A and Cluster B prerequisite skills overlap, the skill(s) will be covered in only one section.

Getting Started
- Allow students time to look over the two pages of the assessment. Point out the labels that identify the skills covered.
- Have students find math vocabulary terms used in the assessment. List vocabulary terms on the board as students identify them. If necessary, review the meanings of all essential math vocabulary.

Introducing the Assessment
- Explain to students that these pages will help you know if they are ready to start a new chapter in their math textbooks.
- Students who have transferred from another school may not have been introduced to some of these skills. Encourage students to do their best and assure them you will help them learn any needed skills.

Cluster A Challenge
Those students who demonstrate mastery of the skills on this page will not need to use the reteaching worksheets. Instead, these students can do the Cluster A Challenge found on page 126.

115C Grade 1, Chapter 10, Cluster A

Name_____

CHAPTER 10 What Do I Need To Know?

Count to 100
Write how many.

1. [grid of faces]   __31__ 😊

Same Number
Write how many.
Circle the same numbers.

2. [groups of faces]   __21__ 😊   __14__ 😊   __14__ 😊

Before and After
Write who comes before and after.

3. Before Emma __Sara__
4. After Mike __Wendy__

Sara Emma Mike Wendy

115A Use with Grade 1, Chapter 10, Cluster A

CLUSTER A PREREQUISITE SKILLS

The skills listed in this chart are those identified as major prerequisite skills for students' success in the lessons in Cluster A of the chapter. Each skill is covered by one or more assessment items as shown in the middle column. The right column provides the page numbers for the lessons in this book that reteach the Cluster A prerequisite skills.

Skill Name	Assessment Items	Lesson Pages
Count to 100	1	116-117
Same Number	2	118
Before and After	3-4	119

Name _____

Ordinals

Write the name of the person.

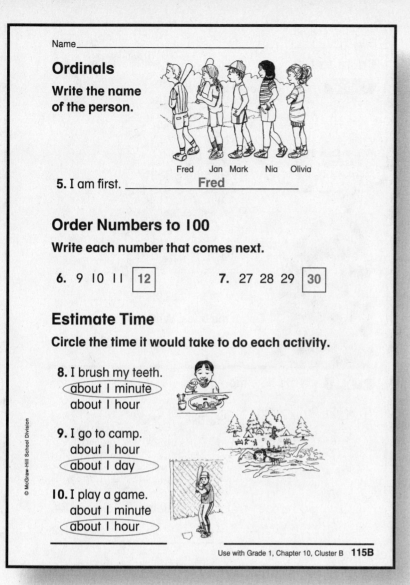

Fred Jan Mark Nia Olivia

5. I am first. _____Fred_____

Order Numbers to 100

Write each number that comes next.

6. 9 10 11 [12]   7. 27 28 29 [30]

Estimate Time

Circle the time it would take to do each activity.

8. I brush my teeth.
   (about 1 minute)
   about 1 hour

9. I go to camp.
   about 1 hour
   (about 1 day)

10. I play a game.
    about 1 minute
    (about 1 hour)

Use with Grade 1, Chapter 10, Cluster B **115B**

CHAPTER 10 PRE-CHAPTER ASSESSMENT

Alternative Assessment Strategies

- Oral administration of the assessment is appropriate for younger students or those whose native language is not English. Read the skills title and directions one section at a time. Check students' understanding by asking them to tell you how they will do the first exercise in the group.

- For some skill types you may wish to use group administration. In this technique, a small group or pair of students complete the assessment together. Through their discussion, you will be able to decide if supplementary reteaching materials are needed.

Intervention Materials

If students are not successful with the prerequisite skills assessed on these pages, reteaching lessons have been created to help them make the transition into the chapter.

Item correlation charts showing the skills lessons suitable for reteaching the prerequisite skills are found beneath the reproductions of each page of the assessment.

CLUSTER B PREREQUISITE SKILLS

The skills listed in this chart are those identified as major prerequisite skills for students' success in the lessons in Cluster B of the chapter. Each skill is covered by one or more assessment items as shown in the middle column. The right column provides the page numbers for the lessons in this book that reteach the Cluster B prerequisite skills.

Skill Name	Assessment Items	Lesson Pages
Ordinals	5	120-121
Order Numbers to 100	6-7	122-123
Estimate Time	8-10	124-125

Cluster B Challenge

Those students who demonstrate mastery of the skills on this page will not need to use the reteaching worksheets. Instead, these students can do the Cluster B Challenge found on page 127.

Grade 1, Chapter 10, Cluster B **115D**

USING THE LESSON

Lesson Goal
- Count tens and ones and the total for numbers through 99.

What the Student Needs to Know
- Count by tens to 100.
- Read and write numbers to 10.
- Identify the tens and ones digits.

Getting Started
- Have students work in small groups. Provide each group with about 100 counters in a jar, bag, or box. Have one student spill out a pile of counters. Say:
- *Make groups of ten with the counters until there are fewer than ten counters left.*

When students have finished, ask: *How many tens do you have? How many ones are left over?*

- Discuss how making groups of ten is a good way to find the total number of counters. Ask: *Why is this better than counting the counters one at a time?* (Answers will vary: You don't lose your place as easily; you can see how many you have at a glance.)
- Have the class count aloud by tens to 100. Repeat, writing each number from 10 through 100 on the chalkboard as the students count.
- Write on the board:
    ____ tens ____ ones
  Have a volunteer from each group come up to the board and write in the numbers to show their number of counters. Make sure students write, for example, 3 tens rather than 30 tens.

What Can I Do?
- Read the question and the response. Then read and discuss the example. Have students trace over the dotted lines to show the three groups of tens. Ask: *How many clocks are there in each group?* (10; if students seem at all uncertain, have them count the clocks to confirm this.)

WHAT IF THE STUDENT CAN'T

Count by Tens to 100
- Students work in pairs, using connecting cubes. Students make 10 cube trains with 10 cubes in each. Students use their trains to practice counting by tens. One student says a number of tens; the other student shows the number with the trains and says the number word. Then students switch roles.
- Provide the student with tens models. Write multiples of tens from 10 to 90 on the chalkboard. Point to number and have the student show you that many tens.

Read and Write Numbers to 10
- Have the student clap ten times as he or she counts to 10. Then write the word name for each number on the chalkboard. As you read each word, have the student write the corresponding number.
- Use number cards with the numbers 0 through 10. As you randomly call out numbers, the student holds up the card.

Name _____

Power Practice • Write how many.

3. [clocks] __3__ tens __1__ one = __31__

4. [clocks] __1__ ten __8__ ones = __18__

5. [clocks] __1__ ten __2__ ones = __12__

6. [clocks] __1__ ten __4__ ones = __14__

7. [clocks] __2__ tens __5__ ones = __25__

8. [clocks] __2__ tens __0__ ones = __20__

Grade 1, Chapter 10, Cluster A 117

WHAT IF THE STUDENT CAN'T

Identify the Tens and Ones Digits
- Use number cards for 20–50. Place the card for 32 on the chalkboard tray. Ask: *What is this number?* (thirty-two) *What does the 3 stand for?* (3 tens.) *What does the 2 stand for?* (2 ones.)

Complete the Power Practice
- Have the student redo any incorrect exercises. As an alternative, the students can use crayons or markers. He or she amy color 10 clocks red, then 10 clocks yellow, and so on. There are at most 3 tens in any exercise, so 3 different colors are needed. Then have the student write the correct answers on the lines.

USING THE LESSON

- *How many tens are there?* (3) Have students trace over the 3 on the writing line. *How many ones are there?* (0 or none) Have students trace the zero.

- *How can we write one number to show 3 tens and 0 ones?* (Students should suggest writing 30. Have them trace over the 30 on the final writing line to complete the example problem.)

Try It
Read the directions aloud. Do the first exercise with students, showing them the steps:

- Circle as many groups of 10 as you can.
- Count the groups of ten and write that number for the tens.
- Count the ones left over and write that number for the ones.
- Write the 2-digit number that shows how many in all.
- Have students complete exercise 2. Check their answers.

Power Practice
- Have students complete the practice items. Remind them to begin by circling groups of ten. Point out that in some exercises, the rows have 5 clocks and in other exercises have 6. So, students cannot assume that two rows equal 10.

- After students have completed the practice items, check their work.

Grade 1, Chapter 10, Cluster A **117**

USING THE LESSON

Lesson Goal
- Identify two groups that have the same number.

What the Student Needs to Know
- Count objects to 10.
- Use one-to-one matching to compare groups.

Getting Started
- Have students work in small groups. Provide each student with 10 counters. Each student displays a number of counters less than 10. Say:
- *You have three groups of counters. Do any two groups have the same number? How can you find out?* (Count each group; match the counters one-to-one.)

What Can I Do?
Read the question and the response. Then read and discuss the example. Ask:
- *How many are there in each group? Count and write the numbers.* Allow students time to do this, then say:
- *Two groups have the same number. What numbers are the same?* (the two 10s) Direct students to circle the 10s.

Try It
Read the directions aloud. Check that students understand the steps for the exercise:
- Count the objects in each group and write the number.
- Find the two numbers that are the same and circle them.

Power Practice
- Have the students complete the practice items. Then review each answer.

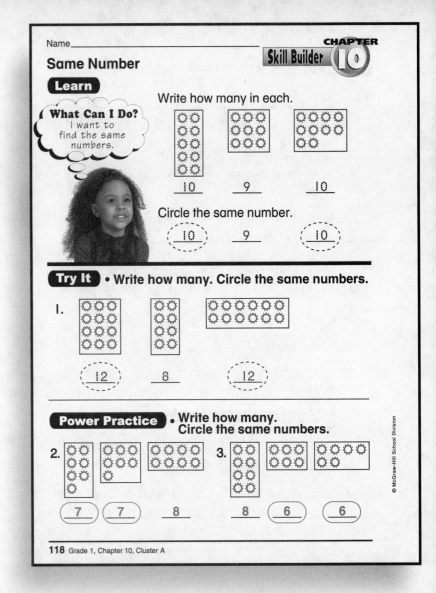

WHAT IF THE STUDENT CAN'T

Count Objects to 10
- Provide number flash cards through 10 and counters. Have students work in pairs. One student draws a number card; the partner displays that number of counters. Students work together to check the number by counting.

Use One-to-One Matching to Compare Groups
- Give two students the same number of counters. Have them show that the numbers are the same by lining up the counters. Then have students each take a handful of counters and compare them by matching them one-to-one. Have students say whether the groups are equal or whether one has more counters and, if so, how many more.

Complete the Power Practice
- Have the student redo any incorrect exercises by recounting each group with you. Then have the student write the correct numbers on the lines and circle the two that are the same.

Name_____

Before and After

Skill Builder CHAPTER 10

Learn

What Can I Do? I want to find who comes before and after.

Who is before Tom?
Look ← to find the one before.

Jean Tom Pat Greg

Jean is before Tom.

Try It • Write who comes before or after.

Greg Jean Pat Tom

1. After Pat __Tom__
2. Before Jean __Greg__

Power Practice • Write who comes before or after.

Jean Pat Greg Tom

3. After Greg __Tom__
4. Before Pat __Jean__

Grade 1, Chapter 10, Cluster A **119**

WHAT IF THE STUDENT CAN'T

Use *Left* and *Right* to Describe Locations
- Point to various pairs of objects and have the student tell which is on the right and which is on the left.
- Have the student fold a sheet of paper in half vertically and label the sides *left* and *right*. Have students follow your directions to place various objects on the left or right side of the paper.

Use *First* and *Last* to Describe Positions of People in Line
- Have the student line up 3 to 5 different geometric shapes and tell which one is first and which is last.
- Students work in pairs using toy cars. They set up a finish line and then put 3 or more cars in a race. Have them tell which car is first and which is last.

Complete the Power Practice
- Have the student redo any incorrect exercises by saying each one aloud and pointing to the appropriate figure. Then have the student write the correct name on the line.

USING THE LESSON

Lesson Goal
- Use *before* and *after* to describe positions of people in line.

What the Student Needs to Know
- Use *left* and *right* to describe locations.
- Use *first* and *last* to describe positions of people in line.

Getting Started
- Give a sheet of paper to each of four students. Have the students line up in front of the room. Ask: *Who is first in line? Who is last in line?* Point to one of the students in the line and ask: *Who comes before this student? Who comes after?*
- Write *before* and *after* on the chalkboard. Read the words and ask students to use them in sentences.

What Can I Do?
- Read the question and the response. Have students look at the four people lined up at the top of the page. Point out that the person on the left side is first in line.
- Read the names of the four people in line. Then ask: *Who comes before Tom in the line?* (Jean)

Try It
- Read the directions aloud. Do exercise 1 with the students. Read the names of the four students standing in line to get their books. Ask: *Which person comes after Pat?* (Tom.) Have students complete exercise 2. Tell students that this time they will look for the person who comes *before* another person.

Power Practice
- Read the names of the four students standing in line. Check that all students know who comes first. Then have them complete the practice items.

Grade 1, Chapter 10, Cluster A **119**

USING THE LESSON

Lesson Goal
- Use ordinal numbers through tenth to describe position.

What the Student Needs to Know
- Use *first, next,* and *last* to describe position.
- Use *before* and *after* to describe relative position.
- Count to 10.

Getting Started
- Have 5 students come up and stand in a line facing left at the front of the class. Ask: *Who is first in line?* Repeat for *second* through *fifth.* Have a new group of five students come up. Ask about the positions in a random order.
- Have ten students make a line facing left at the front of the classroom. Each holds up the appropriate number card 1 through 10. Point to the seventh student and ask: *How can we tell what place this student has?* (number 7; seventh) Write *seventh* on the chalkboard and underline the number word *seven.* Remind students that words such as this show position or location.

What Can I Do?
- Read the question and the response. Then read and discuss the example. Have students look at the line of children. Read each child's name as students point to the name word.
- Review words used for position by asking questions such as the following: *Who is first? Who comes next? Who is last? Who is before Ann? Who comes after Tim?*
- Ask: *How do you decide which one is fourth?* (Start at the left and count to 4.) *What is the answer to the question?* (Tim.)
- Go over the words *first* through *fifth.* Have students point to each word as you read it.

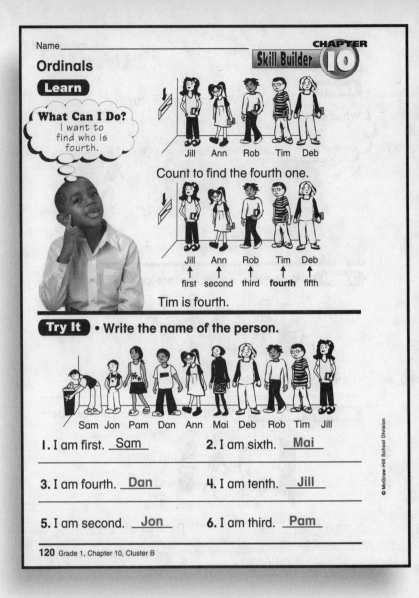

WHAT IF THE STUDENT CAN'T

Use *First, Next,* and *Last* to Describe Position
- Draw a ticket booth on the left side of the chalkboard. Have students come and draw three stick figures to show people in line. Give each figure a name and write the name above it. Have students talk about the order of the people waiting. Ask who is *first, next, last.*
- Have students work in pairs using dolls or other toys. Students line up the figures. Give the pair three cards with the words: *first, next, last.* Students put each card next to the appropriate figure.

Use *Before* and *After* to Describe Relative Position
- Have students work in pairs to create a large drawing of a car or boat race. Students use different colors or numbers to identify the cars or boats. When students' picture is completed, have them write sentences about the picture using the words *before* and *after.*

Name_____

Power Practice • Write the name of the person.

Pam Rob Mai Sam Dan Deb Tim Ann Jill Jon

7. I am second.
   Rob

8. I am sixth.
   Deb

9. I am seventh.
   Tim

10. I am eighth.
    Ann

11. I am ninth.
    Jill

12. I am first.
    Pam

13. I am fifth.
    Dan

14. I am third.
    Mai

Grade 1, Chapter 10, Cluster B **121**

USING THE LESSON

Try It
- Read the directions and do the first exercise with students. Check that all students know what to do by asking: *Who is the first person in line?* (Sam.) Have students write the name on the line.
- Have students complete exercises 2–6. Check their answers.

Power Practice
- Review the names of the students waiting in line. Ask students to point to each name as you read it. Remind them that the line starts at the left.
- Have students complete the practice exercises. Check students' work.

Learn with Partners & Parents
- Use ten dolls, action figures, or toy cars. Line up the figures as if they were waiting their turn. Ask the student to suggest a situation; for example, the toys are waiting to buy ice cream or get tickets for a movie. Hand the student number cards for 1 through 10 and have the student place the cards by each figure. Then ask questions such as "Who is second?" and "Who is tenth?"

WHAT IF THE STUDENT CAN'T

Count to 10
- Students work in pairs using dot cards for 1 through 10. One student draws a card; the partner counts the dots and says the number. Then students switch roles.
- Write the numbers 1 through 10 on the board. Have pairs of students create a number story using a familiar location such as a zoo or park. For example, one student might say, "I went to the park and I saw 1 squirrel." The other student continues with, "I went to the park and I saw 2 park benches." Students take turns making number statements until they reach 10.

Complete the Power Practice
- Have the student rework any incorrect exercises. For each one, ask the student to read the sentence, point to the picture of the person in line, and then write the names.
- Some have trouble reading the words for the ordinal numbers. Have these students write a little number next to each word as a clue to its meaning.

Grade 1, Chapter 10, Cluster B **121**

USING THE LESSON

Lesson Goal
- Write the number that comes just before or just after a number to 100; write a number that comes between two numbers.

What the Student Needs to Know
- Read and write 2-digit numbers.
- Count on from any 2 two-digit number.
- Count back from any 2-digit number.

Getting Started
- Write the numbers from 1 through 10 on the chalkboard. Point to a number and ask: *What number comes just before this number? What number comes just after this number?*
- Write the terms *just before* and *just after* on the chalkboard. Have students use the numbers 1 through 10 to make up statements that use with the two terms. For example, a student might say, "The number 3 comes just before 4."

What Can I Do?
Read the question and the response. Have students point to the 34 on the chart. Ask:
- *What number comes just before 34?* (33) *How can you find this number?* (Count back starting at 34.)
- *What number comes just after 34?* (35) *How can you find this number?* (Count on starting at 34.)
- Have students trace over the dotted numbers at the bottom of the example.

Skill Builder CHAPTER 10

Name_____

Order Numbers to 100

Learn

1	2	3	4	5	6	7	8	9	10
11	12	13	14	15	16	17	18	19	20
21	22	23	24	25	26	27	28	29	30
31	32	33	34	35	36	37	38	39	40
41	42	43	44	45	46	47	48	49	50

What Can I Do? I want to write the numbers that come just before and just after a number.

Count back to find the number just before 34.

33 is just before 34.

Count back to find the number just before 34.

35 is just after 34.

___33___ 34 ___35___

Try It • Write the number that comes just before.

1. __20__ 21   2. __35__ 36   3. __49__ 50

Write the number that comes just after.

4. 15 __16__   5. 30 __31__   6. 44 __45__

Write the number that comes between.

7. 36 __37__ 38   8. 39 __40__ 41

122 Grade 1, Chapter 10, Cluster B

WHAT IF THE STUDENT CAN'T

Read and Write 2-Digit Numbers
- Have students work in pairs using tens and ones models. One student shows some tens and ones; the partner counts the tens, counts the ones, and then writes the number. Have students switch roles.
- Have students work in pairs using a set of 2-digit number cards. Students take turns drawing a card, saying the number, and then writing it.

Count On from Any 2-Digit Number
- Write the term *count on* on the chalkboard and explain that it means to count forward by ones. As an example, start at 5 and count up to 7.
- Students work in pairs using a set of 2-digit number cards. They turn the cards over and mix them up. One student draws a card and says the number. The partner counts on two numbers. Students switch roles.

Name_____

Power Practice • Write the number that comes just before.

9. _27_ 28   10. _39_ 40

1	2	3	4	5	6	7	8	9	10
11	12	13	14	15	16	17	18	19	20
21	22	23	24	25	26	27	28	29	30
31	32	33	34	35	36	37	38	39	40
41	42	43	44	45	46	47	48	49	50
51	52	53	54	55	56	57	58	59	60

11. _13_ 14   12. _22_ 23

13. _30_ 31   14. _12_ 13   15. _25_ 26

Write the number that comes just after.

16. 54 _55_   17. 40 _41_   18. 10 _11_

19. 22 _23_   20. 46 _47_   21. 59 _60_

Write the number that comes between.

22. 50 _51_ 52   23. 28 _29_ 30

24. 42 _43_ 44   25. 20 _21_ 22

Grade 1, Chapter 10, Cluster B **123**

WHAT IF THE STUDENT CAN'T

Count Back from Any 2-Digit Number

- Write the term *count back* on the chalkboard and explain that it means to count backward by ones. As an example, start at 10 and count back to 5.
- Students work in pairs using a set of 2-digit number cards. One student draws a card and says the number. The partner counts back one number.

Complete the Power Practice

- Have the student redo any incorrect exercises. Ask the student to find the given number on the chart and then find the number that comes just before or just after.
- When the student is trying to locate a number that "comes between," have the student point to the two numbers on the chart and then identify the one that comes between.
* Have the student write the correct answer on the line.

USING THE LESSON

Try It

Read the first direction aloud. Ask: *What number is shown in the first exercise?* (21) Have students find 21 in the chart. Ask: *Where is the number that comes just before?* (Students should find 20 at the top right of the row above.) Have students complete the first row of exercises.

- Read the second direction for students. Have a volunteer explain how to use the chart to do the first exercise in this row. Have students complete the row.

- Have students look at Exercise 7. Ask them how this exercise is different. (There are two numbers printed, with the writing line in between the two numbers.)

- Read the third direction and point out that students find the number that comes just after the number on the left. When they are finished, the three numbers should be in order.

Power Practice

- Have the students look over the page. Ask: *What does the chart at the top show?* (the numbers 1 through 60)

- *How can you use the chart to find a number that comes just before?* (Find the number and then look to the right.) *What happens when you get to the end of a row?* (You start at the left of the next row.)

- Point out that there is a new set of directions after Exercise 15. Have a student read these directions.

- Ask: *Where does the next direction begin?* (after Exercise 21, or before Exercise 22) Have a student read these directions.

Grade 1, Chapter 10, Cluster B **123**

USING THE LESSON

Lesson Goal
- Choose time estimates for everyday activities.

What the Student Needs to Know
- Identify *minute, hour,* and *day* as units to measure time.
- Identify the number of minutes and hours on a clock.
- Tell which of two events will take more time.

Getting Started
- Say: *Name an activity you did this morning.* (Students may describe eating breakfast, riding on the bus, and so on.) Make a chart on the chalkboard with columns labeled: *less time, about the same time, more time.* Choose one of the activities named by students and write it on the chalkboard. Have students use this activity for comparison and name other activities to go in the chart.
- Have students look at the classroom clock. Ask: *What do the numbers on the clock show?* (hours and minutes) *How many hours are shown on the clock?* (12)
- *Which is longer, 1 hour or 1 minute?* (1 hour) *How many minutes are there in an hour?* (60)
- Write on the chalkboard: *about 1 minute, about 1 hour.* Say: *If it is cold, you will put on a jacket. Which is the better estimate of the time it takes, 1 minute or 1 hour?* (1 minute)
- Have students look at the classroom calendar. Ask: *Which is longer, 1 hour or 1 day?* (1 day) *How many days are there in a week?* (7) *How many days in a week do you go to school?* (5)
- Write on the chalkboard: *about 1 day, about 1 hour.* Say: *A person takes a dog for a walk. Which is the better estimate of the time it takes, 1 day or 1 hour?* (1 hour)

What Can I Do?
Read the question and the response, then read and discuss the example. Have students look at

WHAT IF THE STUDENT CAN'T

Identify *Minute, Hour,* and *Day* As Units to Measure Time
- Hold up a clock and a ruler. Ask: *What do we measure with each of these tool?* (time with the clock; distance or length with the ruler) *We measure length in inches or centimeters. How do we measure time?* (minutes, hours, second, days, months, and so on) Point out that another tool used to measure time is a calendar.
- Show students a kitchen timer, a clock, and a calendar. Discuss how each is used to measure time and the units used for each.

Identify the Number of Minutes and Hours on a Clock
- Use the demonstration clock. Point out the two hands on the clock and that the hands are different lengths. Ask: *Which hand points to the hour?* (the shorter hand) *Which hand points to the minutes?* (the longer hand)
- Show students how the hour hand moves from one number to the next each hour as the minute hand moves around the clock. *How many hours are shown on the clock?* (12 because there are 12 numbers)

Name_____

Power Practice • Circle the time it would take to do each activity.

3. I read a book.

about 1 minute
(about 1 hour)

4. I put on my socks.

(about 1 minute)
about 1 hour

5. I go to Grandma's house.

about 1 minute
(about 1 day)

6. I wash my face.

(about 1 minute)
about 1 hour

7. I fill a glass.

(about 1 minute)
about 1 hour

8. I watch a TV show.

(about 1 hour)
about 1 day

Grade 1, Chapter 10, Cluster B **125**

WHAT IF THE STUDENT CAN'T

- Model for students how the minute hand makes one rotation in an hour. Have them count by 5s to see that there are 60 minutes in one hour.

Tell Which of Two Events Will Take More Time

- Write on the chalkboard: *Play a game of baseball. Build a house.* Ask students which activity takes longer and why. (Building a house will take weeks or months. A game of baseball takes less than one day.) Repeat with other pairs with obvious time differences.
- Students work in small groups. One student begins by naming a rather short activity, for example, eating an apple. Each other student names an activity that takes more time. Repeat until each student has started a round.

Complete the Power Practice

- Have the student redo any incorrect exercises. For each one, have the student explain his or her reasoning so that you can resolve the error. Then have the student circle the correct answer.

USING THE LESSON

each activity in turn. For each of them, ask: *About how long will this take, 1 minute, 1 hour, or 1 day?*

- Read each of the three time estimates and have students point to the picture that goes with each one.
- Remind students that the word *about* is used for many kinds of estimates. Use examples familiar to students such as "about 1 foot long" and "about 1 gallon."

Try It

- Read the directions aloud and do the first exercises with students. Ask: *How many answer choices are there?* (two) *Which is the better time estimate for riding a bike?* (One hour.) Have students trace the dotted line circle around the answer.
- Have students complete exercise 2. Check their answers.

Power Practice

- Read the directions and the exercises aloud with students.
- Then have them complete the practice items. Review answers with students.

Grade 1, Chapter 10, Cluster B **125**

CHALLENGE

Lesson Goal

- Solve number puzzles using clues and logical reasoning.

Introducing the Challenge

Display a large hundred chart:

1	2	3	4	5	6	7	8	9	10
11	12	13	14	15	16	17	18	19	20
21	22	23	24	25	26	27	28	29	30
31	32	33	34	35	36	37	38	39	40
41	42	43	44	45	46	47	48	49	50
51	52	53	54	55	56	57	58	59	60
61	62	63	64	65	66	67	68	69	70
71	72	73	74	75	76	77	78	79	80
81	82	83	84	85	86	87	88	89	90
91	92	93	94	95	96	97	98	99	100

Ask questions such as:

- *What numbers are between 24 and 28?* (25, 26, 27)
- *Look at the numbers in the third row. In 21, the sum of the two digits is 3. For which other number in the row is this true?* (30)

Using the Challenge

- Read the directions. Help students get started with Problem 1 by reading the first two clues. Sketch a number line to show how these clues narrow the possible answers to 31, 32, 33, 34.
- Students may need help with the clue, "If you add my numbers, the sum is 4." If necessary, explain that this means to add the two digits in the number.
- Students may do the Challenge independently, in pairs or small groups. Provide them with a hundred chart.
- When students have completed the Challenge, they may enjoy creating mystery numbers of their own.

Name____________________

CHAPTER 12 CHALLENGE

Mystery Numbers

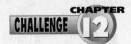

Read each clue.
Find each mystery number.

1. I am after 30.
I am before 35.
I am **not** between 32 and 34.
If you add my numbers, the sum is 4.
What number am I? 31

2. I am after 20.
I am before 40.
One of my numbers is zero.
What number am I? 30

3. I am after 10.
I am before 20.
I am **not** after 13.
I am **not** 11.
What number am I? 12

4. I am between 20 and 30.
If you add my numbers, the sum is 11.
What number am I? 29

126 Grade 1, Chapter 10, Cluster A

Name_____

Picturing Time

CHAPTER 10 CHALLENGE

Draw a picture of something you do for each time that is shown.
Write a sentence about each picture.

About 1 minute

[ ]

Answers may vary. Check children's work.

_____

About 1 hour

[ ]

Answers may vary. Check children's work.

_____

Grade 1, Chapter 10, Cluster B **127**

CHALLENGE

Lesson Goal
- Describe and illustrate activities that take about 1 minute and about 1 hour.

Introducing the Challenge
- Write *hour* and *minute* on the chalkboard. Ask: *What do these measure?* (time) *Which one is longer? How do you know?* (An hour is longer; there are 60 minutes in 1 hour.)
- One hour before lunch or recess, have students look at the classroom clock. Write the time on the chalkboard. When 1 hour has passed, point this out to students. Ask: *What can you do that takes 1 hour?* (Accept all plausible answers.)
- Have students clap every second for 1 minute. Ask: *What can you do that takes 1 minute?* (Accept all plausible answers.)

Using the Challenge
- Read the directions. Check that students understand by asking: *What will you draw in the first box?* (An activity that takes about 1 minute to complete.) *What are the writing lines for?* (to describe the activity in a sentence) *What will you draw in the second box?* (An activity that takes about 1 hour to complete.)
- If students need help with the sentences that describe the activities, suggest that they begin with the pronoun *I*. For example, "I brush my teeth every night." and "I like to play with the dog."

Grade 1, Chapter 10, Cluster B **127**

Name_____

CHAPTER 11 — What Do I Need To Know?

Count to 100

Write how many.

1. [cube train of 10] ____ [single cube]

2. [three cube trains of 10] [two cube trains of 10] ____ [single cube]

Longer or Shorter

Circle the one that is longer.

3.

4.

127A Use with Grade 1, Chapter 11, Cluster A

Name _____

More Than, Less Than, About the Same As

**Compare the amount of water in the bottles.
Circle the one that has more.**

5.   6.

Circle the bottle that has about the same amount.

7.

**Compare the bottles on the scale.
Circle the side that has less.**

8. 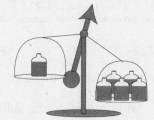   9.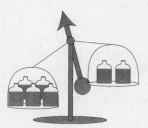

Circle the one that is heavier.

10.

Use with Grade 1, Chapter 11, Cluster B **127B**

CHAPTER 11 PRE-CHAPTER ASSESSMENT

Assessment Goal
This two-page assessment covers skills identified as necessary for success in Chapter 11 Measurement. The first page assesses the major prerequisite skills for Cluster A. The second page assesses the major prerequisite skills for Cluster B. When the Cluster A and Cluster B prerequisite skills overlap, the skill(s) will be covered in only one section.

Getting Started
- Allow students time to look over the two pages of the assessment. Point out the labels that identify the skills covered.
- Have students find math vocabulary terms used in the assessment. List vocabulary terms on the board as students identify them. If necessary, review the meanings of all essential math vocabulary.

Introducing the Assessment
- Explain to students that these pages will help you know if they are ready to start a new chapter in their math textbooks.
- Students who have transferred from another school may not have been introduced to some of these skills. Encourage students to do their best and assure them you will help them learn any needed skills.

Cluster A Challenge
Those students who demonstrate mastery of the skills on this page will not need to use the reteaching worksheets. Instead, these students can do the Cluster A Challenge found on page 136.

127C Grade 1, Chapter 11, Cluster A

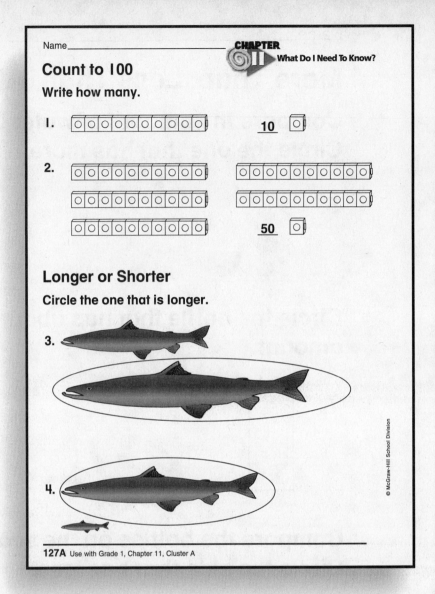

CLUSTER A PREREQUISITE SKILLS

The skills listed in this chart are those identified as major prerequisite skills for students' success in the lessons in Cluster A of the chapter. Each skill is covered by one or more assessment items as shown in the middle column. The right column provides the page numbers for the lessons in this book that reteach the Cluster A prerequisite skills.

Skill Name	Assessment Items	Lesson Pages
Count to 100	1-2	128-129
Longer or Shorter	3-4	130-131

Name_____

More Than, Less Than, About the Same As

Compare the amount of water in the bottles.
Circle the one that has more.

5.   6.

Circle the bottle that has about the same amount.

7.

Compare the bottles on the scale.
Circle the side that has less.

8.   9.

Circle the one that is heavier.

10.

Use with Grade 1, Chapter 11, Cluster B **127B**

CLUSTER B PREREQUISITE SKILLS

The skills listed in this chart are those identified as major prerequisite skills for students' success in the lessons in Cluster B of the chapter. Each skill is covered by one or more assessment items as shown in the middle column. The right column provides the page numbers for the lessons in this book that reteach the Cluster B prerequisite skills.

Skill Name	Assessment Items	Lesson Pages
More Than, Less Than, About the Same As	5-10	132-135

CHAPTER 11 PRE-CHAPTER ASSESSMENT

Alternative Assessment Strategies

- Oral administration of the assessment is appropriate for younger students or those whose native language is not English. Read the skills title and directions one section at a time. Check students' understanding by asking them to tell you how they will do the first exercise in the group.
- For some skill types you may wish to use group administration. In this technique, a small group or pair of students complete the assessment together. Through their discussion, you will be able to decide if supplementary reteaching materials are needed.

Intervention Materials

If students are not successful with the prerequisite skills assessed on these pages, reteaching lessons have been created to help them make the transition into the chapter.

Item correlation charts showing the skills lessons suitable for reteaching the prerequisite skills are found beneath the reproductions of each page of the assessment.

Cluster B Challenge

Those students who demonstrate mastery of the skills on this page will not need to use the reteaching worksheets. Instead, these students can do the Cluster B Challenge found on page 137.

USING THE LESSON

Lesson Goal
- Count the tens and ones in multiples of 10 through 100.

What the Student Needs to Know
- Show 10 with connecting cubes.
- Count to 10.
- Identify the tens and ones digits in numbers.

Getting Started
- Review the concept of zero. Draw a fish bowl on the chalkboard and draw 3 fish in it. Ask students how many fish are in the bowl. Erase the fish and ask: *How many fish are in the bowl now?* (zero, none, no fish) *What number do we used to show nothing?* (zero)
- Remind students that numbers greater than 9 are written with two digits. The first digit shows the tens; the second digit shows the ones. Write on the chalkboard:

  3 tens 2 ones = 32

  Have students use counters or connecting cubes to show this number.
- Direct students to take away 2 ones so they have 3 tens left. Ask: *How many tens are in this number?* (three tens) *How many ones?* (zero ones) *How do we write this number?* (30) Emphasize that the 3 shows the tens; the 0 shows the ones.

What Can I Do?
Read the question and the response. Then read and discuss the example. Ask:
- *How many cubes are in each train?* (10) *How many trains are there?* (4) *How many tens is that?* (4) *Are there any single cubes?* (no) *What number shows there are no ones?* (zero) Have students trace over the numbers to complete the example.

Try It
Read the directions aloud and do exercise 1 with students.

128 Grade 1, Chapter 11, Cluster A

WHAT IF THE STUDENT CAN'T

Show 10 With Connecting Cubes
- Have students work in pairs. Give each pair about 25 connecting cubes. Have each student make a cube train with 10 cubes in it.
- Students may find it easier to count the cubes in a train if the cubes are not all the same color. Have students use two colors and alternate them; for example, a student might alternate yellow cubes with red cubes. Or, a student could show counting by 2s by connecting 2 reds, 2 yellows, 2 reds, 2 yellows, 2 reds.

Count to 10
- Write the numbers 1 through 10 on the chalkboard. Have the student count aloud to 10 and point to each number as he or she says it.
- Provide dot cards and number cards. Students work in pairs to match up the cards.
- Students work in pairs with number flash cards and counters. One student draws a card; the other student shows that number with counters. Then students switch roles.

Name _____

Power Practice • Write how many.

3. [cubes]
   _5_ tens _0_ ones = _50_

4. [cubes]
   _7_ tens _0_ ones = _70_

5. [cubes]
   _9_ tens _0_ ones = _90_

Grade 1, Chapter 11, Cluster A **129**

USING THE LESSON

- Ask the students to count the tens and write the number before the word "tens."
- Count the ones and write the number before the word "ones."
- Write the number that shows the number of cubes.
- Have students complete exercise 2. Check their answers.

Power Practice
- Read the directions aloud and check that students understand the steps. Ask: *What will you show in each exercise?* (the number of tens, the number of ones, and the number of cubes in all)
- Have students complete the practice exercises. Then check their answers.

WHAT IF THE STUDENT CAN'T

Identify the Tens and Ones Digits in Numbers
- Use a set of number cards—from 10–50. Hold up a number card. Have the student read the number and then tell you the number of tens and ones.
- Have the student fold a sheet of paper in half vertically. Write the words *tens* and *ones* on the chalkboard. Have the student label the two sides of the paper. Call out various two-digit numbers and have the student put the appropriate place-value models on each side of the paper.

Complete the Power Practice
- Have the student use tens models to show the numbers for each incorrect exercise. Then ask the student to write the correct numbers on the lines.

USING THE LESSON

Lesson Goal
- Identify the longer or shorter of two objects.

What the Student Needs to Know
- Use *same* and *different* to describe pairs of objects.
- Tell whether or not two objects are the same length.
- Use *top* and *bottom* to describe relative position.

Getting Started
- Have students work in small groups. Give each student a pencil, pen, crayon, or marker. The students in each group work together to arrange the items in order from shortest (at the top) to longest (at the bottom). Show students how to draw a vertical line so they can be sure the left ends of the items are lined up.
- Have the students turn the pencils and markers so they point up instead of to the right. Point out that the pencils are now arranged from shortest to tallest. Contrast the use of the comparison words *longer, taller; longest; tallest* by having students use them in sentences to describe things in the classroom.

What Can I Do?
- Read the question and the response. Have students look the two fish at the top of the page. Ask: *Are the two fish the same?* (no) *How are they different?* (One is longer than the other.) *Which fish is longer, the top one or the bottom one?* (the top fish) Have students trace the dotted circle to show that the top fish is the longer one in the pair.
- Point out that the two fish are lined up on the left sides. Ask: *Why is it important to line up the left sides of the fish?* (You might not be able to tell which fish was longer if they weren't lined up.)

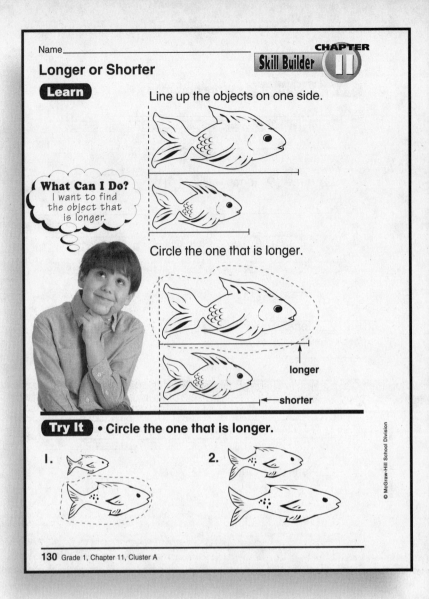

WHAT IF THE STUDENT CAN'T

Use *Same* and *Different* to Describe Pairs of Objects
- Give a variety of attribute blocks. Have students match the ones that are the same.
- Students work in pairs with pattern blocks. One student chooses two blocks; the partner tells whether they are the same or different. If the blocks are different, the student describes the differences in color, shape, or size.

Tell Whether or Not Two Objects Are the Same Length
- Show students two crayons that are identical except for color. Ask: *Is one crayon longer than the other? How do you know?* (Students line up the ends to compare lengths.) Point out that the crayons are the same length. Repeat with two crayons of different lengths.

Name_____

Power Practice • Circle the one that is longer.

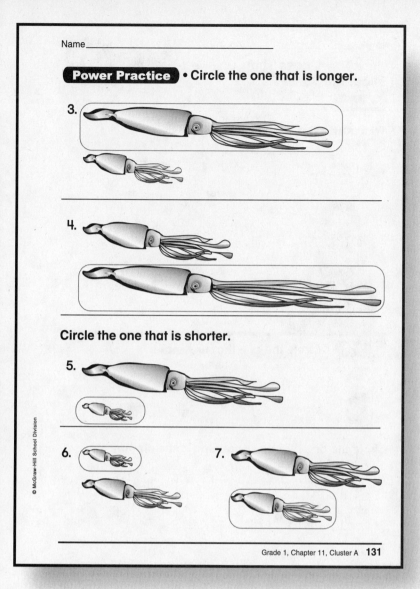

3.

4.

Circle the one that is shorter.

5.

6.   7.

Grade 1, Chapter 11, Cluster A **131**

WHAT IF THE STUDENT CAN'T

Use *Top* and *Bottom* to Describe Relative Position

- Have students fold a piece of paper in half horizontally. Ask them to draw a fish in one half and a flower in the other half. Direct students to turn the paper so that the fish is at the top. Check that students understand that *top* means the side away from them. Then have students turn the paper so that the flower is at the top. The papers can also be used for practice with *right* and *left*.

Complete the Power Practice

- Have the student rework any incorrect exercise. The student can use cube trains to measure the length of each object.
- Then have the student circle the longer or shorter object.

USING THE LESSON

Try It

- Read the directions aloud. Check that students understand they are to compare the two fish in each exercise. Have them point to the fish for exercise 1. Ask: *How will you decide which fish is longer?* (Look at the right side.) *How will you show your answer?* (Circle the longer fish.) Have students trace the circle.
- Have students complete exercise 2 on their own. Review their answers.

Power Practice

- Read the first set of directions for exercises 3 and 4 and the second set of directions for exercises 5–7. Be sure students understand what to do.
- Have students complete the practice exercises. Check students answers.

Learn with Partners & Parents

- Have students work with a partner. Use a roll of adding machine tape. Each student tears off a strip of tape. They estimate which strip is longer, then work together to lay out the tapes to compare their lengths.
- Write each of these comparison words on a different piece of paper: *longer, shorter, taller, farther, wider, thinner.* All of these words refer to length of distance. Students create an illustration and write a sentence for each word; for example, "The school is farther from home than the library."

Grade 1, Chapter 11, Cluster A **131**

USING THE LESSON

Lesson Goal
- Compare liquid amounts in identical bottles.

What the Student Needs to Know
- Identify the container that holds more liquid.
- Identify the container that holds less liquid.
- Choose two containers with the same size and shape.

Getting Started
- You will need a variety of clear plastic containers, water, and food coloring for this lesson.
- Hold up two obviously different containers such as a small plastic cup and a large gallon jug. Ask: *If I fill these with colored water, which will hold more?* (the jug)
- *Here is a way to prove that the jug holds more.* Fill the small cup with colored water and then pour it into the jug. Point out how much extra space there is in the jug.
- Set out two identical containers. Ask: *If these are both full, which one holds more?* (They will hold the same.) *If they are not full, can one container hold more than the other?* (Yes, if the containers are not full, one can have more.)
- Using two identical containers, fill one almost completely with colored water. Pour only a little liquid into the other container. Ask: *Which has more liquid? How do you know?* (The one that is almost full has more.)
- Explain to students that, if two containers are the same, you can see which has more by looking at the line of water. The higher the line, the more water. This method *only* works, however, if the containers are identical.

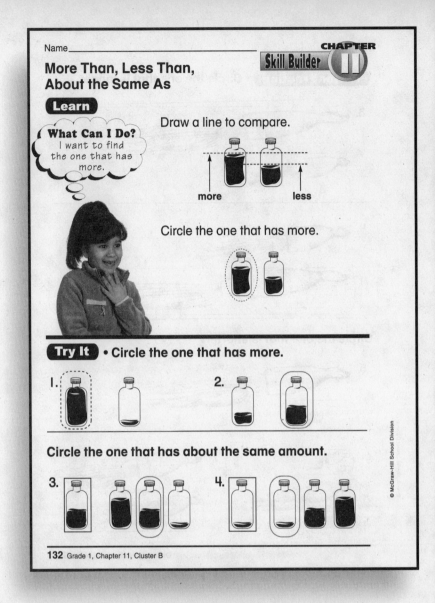

WHAT IF THE STUDENT CAN'T

Identify the Container That Holds More Liquid
- Provide two obviously different plastic containers and colored water. Have the student fill the smaller container with water and then pour the water into the larger container. Pour more water into the larger container to fill it, pointing out that this container holds more water than the smaller container.

Identify the Container That Holds Less Liquid
- Use two obviously different plastic containers, colored water, and a large basin or sink. Have the student fill the larger container with water and then try to pour the water into the smaller container.
- Point out that the smaller container gets filled up and there is still water in the larger container. So, the smaller container holds less liquid than the larger container.

Choose Two Containers with the Same Size and Shape
- Put out three containers, two identical and one obviously different. Have the student tell you which two containers are the same.

Name_____

Power Practice • Circle the one that has more.

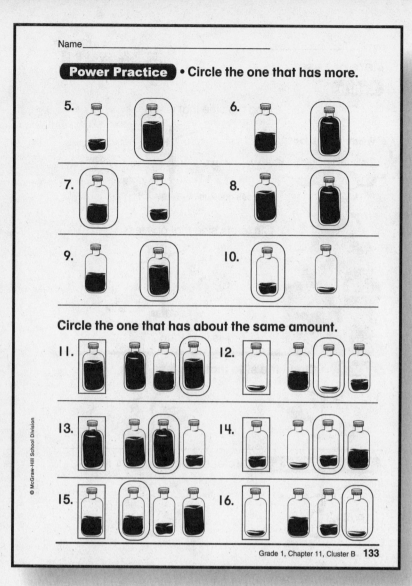

Circle the one that has about the same amount.

Grade 1, Chapter 11, Cluster B **133**

WHAT IF THE STUDENT CAN'T

- Provide a pair of students with about 8 different containers, including two pairs of identical containers. Students work together to find the pairs that are identical. Remind students to use more than just the bottle height. They should also look at the width and the general shape.

Complete the Power Practice
- Have the student rework any incorrect exercises. Ask the student to explain the reason for each answer choice. Then have the student circle the appropriate container.

USING THE LESSON

What Can I Do?
- Read the question and the response. Then read and discuss the example. Ask: *Is either bottle full?* (No.) *In which bottle is the liquid closer to the top?* (The bottle on the left.) *Which bottle has more liquid?* (The bottle on the left.)
- Point out that students show their answer by circling the bottle that has more liquid.

Try It
- Read the first set of directions aloud and do exercise 1 with students. Ask them to look at the two bottles and circle the one with more liquid. Have students do exercise 2 on their own. Check students' answers.
- Read the second set of directions aloud. Be sure students understand that this time they are to circle the bottle that has about the same amount. Have students complete exercises 3 and 4. Check students' answers.

Power Practice
- Read the directions for exercises 5–10 and 11–16. Make sure students understand what they are to do. Have them complete the page. Check their work.

Grade 1, Chapter 11, Cluster B **133**

USING THE LESSON

Lesson Goal
- Compare weights based on identical standard units.

What the Student Needs to Know
- Use a balance scale to compare weights.
- Use *left* and *right* to describe location.

Getting Started
- You will need a balance scale and identical objects to use as weights. Use cubes or any small objects. Put two objects on each side of the scale and show students that the pans balance.
- Ask: *If I take away one from this side, what will happen?* (Accept all conjectures.) Take away one object to show that this makes one side of the scale go up. Ask: *Why does this side go up?* (There is less weight on it now.) *How can I make the two sides balance again?* (Put one object back so there are the same number of objects on both sides.)

What Can I Do?
Read the question and the response. Then read and discuss the example.
- Have students look at the top picture showing the balance scale. Ask: *Why is one side lower than the other?* (It has more weight on it.) *Which side has less weight? How do you know?* (The left side; there are fewer pails of sand on that side.) Have students circle the side with less weight.

Try It
- Read the directions. Check that students know what to do by asking: *Which side of the scale will you circle?* (The one that has less weight.)

Power Practice
- Have the students complete the practice items. When they have finished, ask: *Which side of the scale is lower?* (The one with more pails of sand on it.)

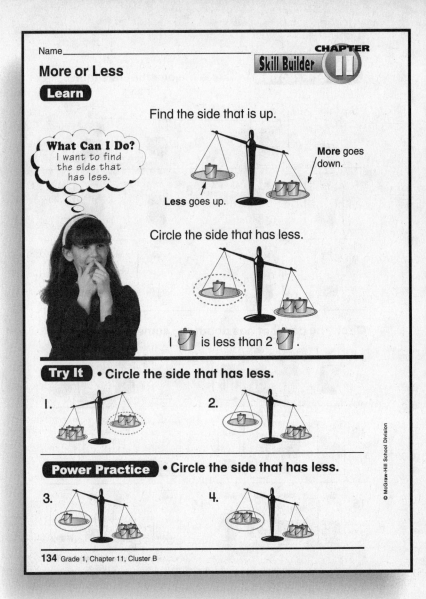

WHAT IF THE STUDENT CAN'T

Use a Balance Scale to Compare Weights
- Have students work in pairs to experiment with a pan balance. Provide various objects for them to put on the pans. After students have experimented for a while, have them hold two different objects, one in each hand, and estimate which has less weight. They use the scale to check their estimates.

Use *Left* and *Right* To Describe Location
- Have the student come to the board and draw a circle. Note which hand the student uses. Draw a vertical line to divide the circle in half. Write "right" on the right side of the circle; "left" on the left. Point out that the right (or left) hand is the hand the student uses to draw. Ask the student to draw something in each part of the circle and describe what he or she drew using *left* and *right*.

Complete the Power Practice
- Have students use an actual pan balance and model each incorrect exercise. Any standard objects can represent pails of sand. Have students complete each item correctly.

USING THE LESSON

Lesson Goal
- Identify the heavier of two objects.

What the Student Needs to Know
- Identify weight as a property of objects.
- Describe how a balance scale compares weights.

Getting Started
- You will need a balance scale and a variety of objects to use as weights. Have a student select two objects, one obviously bigger and heavier than the other. The student puts the objects on the pans of the scale. Ask: *Why does one side of the scale go down?* (That side has more weight on it.) *How does the scale compare two objects?* (It shows which one is heavier.)
- Write the term *heavy* on the chalkboard and have students use it to describe something in the classroom. Then write *heavier* and point out that we use this word to compare two objects. Have students find two things in the classroom to compare using the word *heavier*.

What Can I Do?
- Read the question and the response. Then read and discuss the example. The balance scale shows that the dog is heavier. Have students circle the picture of the dog.

Try It
- Read the directions aloud. Have students identify each object pictured before they begin.
- Do the first exercise with students.

Power Practice
- Have the students complete the practice items.

WHAT IF THE STUDENT CAN'T

Identify Weight As a Property of Objects
- Put something heavy in one of two small boxes. Pad the object with paper so it doesn't rattle. Have the student lift the boxes to compare them. Ask: *What is different about the boxes?* (One is heavier.) Explain that weight is a property all objects have.

Describe How a Balance Scale Compares Weights
- Have students work in pairs to experiment with a pan balance. Provide various objects for them to put on the pans. After students have experimented for a while, have them hold two different objects, one in each hand, and estimate which is heavier. They can then use the scale to check their estimates.

Complete the Power Practice
- Reveiw each incorrect response with the student. Ask the student to explain why he or she chose a given item. Then discuss why one item is heavier than another. Have the student circle the correct answer.

CHALLENGE

Lesson Goal
- Apply length comparisons to everyday objects.

Introducing the Challenge
- Review the terms *longer* and *shorter*. Write each word on the chalkboard and have volunteers use each one in a sentence. Point out that you need to compare two things to use each word.
- Ask students to suggest things that are long in length. Write students' idea on the board in a vertical list. Then, for each item, students must think of something that is even longer. Repeat the activity with a list of short things.

Using the Challenge
- Read the directions. Discuss with students what is meant by "real life." Suggest that they think of objects they see at school, at home, or in other everyday situations.
- Review the pictured objects in the problems, making sure students can identify each one. Point out that the objects are *not* shown in their actual size.
- You may want students to decide ahead of time what they are going to draw. They can describe their ideas to you before they begin so you can discuss them.
- When students have completed the activity, have them share their pictures.

Name_____

CHALLENGE 11

The Long and the Short of It

Draw something in real life that is longer.

1–4 Answers will vary. Check children's drawings.

1. Longer than a [crayon].

2. Longer than a [ruler 12"].

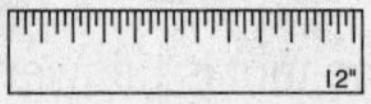

Draw something in real life that is shorter.

3. Shorter than a [door]. 4. Shorter than a [teddy bear].

Name _____

Heavy Duty

CHALLENGE CHAPTER 11

Put the things in order from lightest to heaviest.
Write the letters on the lines to show the order.

1. S A N
   (A) N (S)

2. M E L
   (L) M (E)

3. T C L
   (L) C (T)

Circle the letters of the lightest and heaviest things in each row. Read down the letters of the lightest things. Read down the letters of the heaviest things. Write the message here.

A L L   S E T

Grade 1, Chapter 11, Cluster B **137**

CHALLENGE

Lesson Goal
- Compare weights of objects and solve a code puzzle.

Introducing the Challenge
- Draw on the chalkboard (or use pictures of) a horse, a butterfly, and a bottle of juice. Ask students to identify the letter that each word begins with. Write H, B, and J under the objects.
- Referring to the pictures on the chalkboard, ask: *Do these all weigh the same?* (No.) *Which one is the heaviest?* (the horse) *the lightest?* (the butterfly) *Put the letters in order to show lightest, middle weight, heaviest.* (B, J, H)
- Have students suggest something heavy. Write the name of the object on the chalkboard and label it "heavy." Ask students to name something that is heavier. Finally, ask students to name something that is heavier than the other two, or *heaviest*. Have students use "heaviest" in a sentence. Point out that *heavier* is used to compare one thing with another; *heaviest* is used to compare one thing with two or more other things.

Using the Challenge
- Read the directions. Check that students understand by asking: *For the first problem, which object is the lightest?* (the ant) *What letter will you write on the first writing line?* (A for ant)
- After students have completed the three problems, help them follow the directions at the bottom to complete and solve the puzzle.

Grade 1, Chapter 11, Cluster B **137**

Name_____

CHAPTER 12 — What Do I Need To Know?

Spatial Order

Draw each shape.

1. the shape inside the box   2. the shape outside the box

           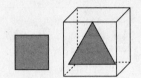

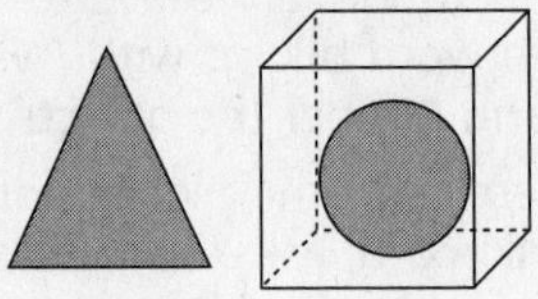

_____   _____

Follow Directions

Draw an X on the third animal.

3.

Sort by Shape

Circle the shape that rolls.

4.   📕

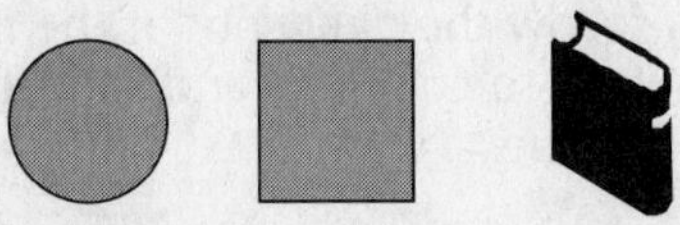

Circle the shapes that are the same.

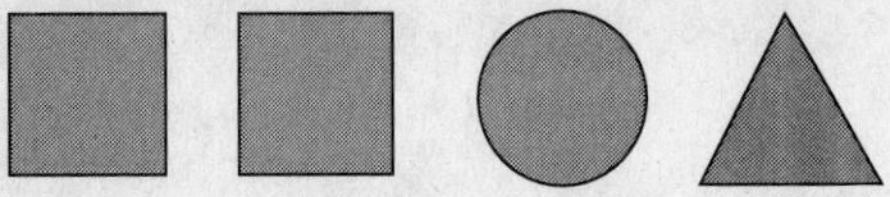

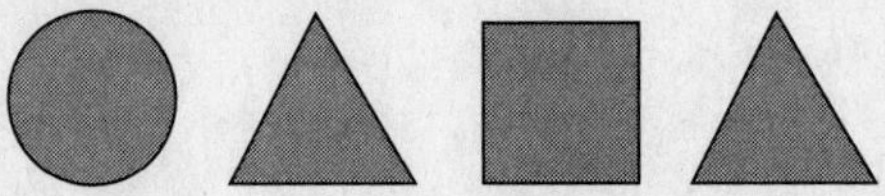

5.   6.

Name_____

Sort by Size

Circle the shape that is bigger than the others.

7.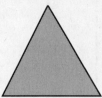

8.

Circle the shapes that are the same size.

9.

10.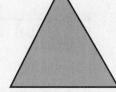

CHAPTER 12 PRE-CHAPTER ASSESSMENT

Assessment Goal

This two-page assessment covers skills identified as necessary for success in Chapter 12 Geometry. The first page assesses the major prerequisite skills for Cluster A. The second page assesses the major prerequisite skills for Cluster B. When the Cluster A and Cluster B prerequisite skills overlap, the skill(s) will be covered in only one section.

Getting Started

- Allow students time to look over the two pages of the assessment. Point out the labels that identify the skills covered.
- Have students find math vocabulary terms used in the assessment. List vocabulary terms on the board as students identify them. If necessary, review the meanings of all essential math vocabulary.

Introducing the Assessment

- Explain to students that these pages will help you know if they are ready to start a new chapter in their math textbooks.
- Students who have transferred from another school may not have been introduced to some of these skills. Encourage students to do their best and assure them you will help them learn any needed skills.

Cluster A Challenge

Those students who demonstrate mastery of the skills on this page will not need to use the reteaching worksheets. Instead, these students can do the Cluster A Challenge found on pages 148-149.

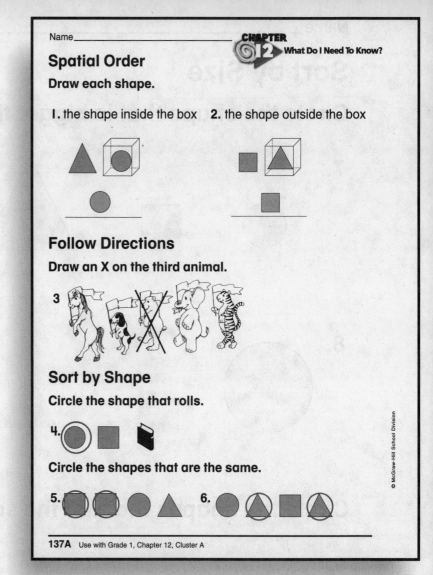

CLUSTER A PREREQUISITE SKILLS

The skills listed in this chart are those identified as major prerequisite skills for students' success in the lessons in Cluster A of the chapter. Each skill is covered by one or more assessment items as shown in the middle column. The right column provides the page numbers for the lessons in this book that reteach the Cluster A prerequisite skills.

Skill Name	Assessment Items	Lesson Pages
Spatial Order	1-2	138-139
Follow Directions	3	140-141
Sort by Shape: Shapes that Roll	4	142-143
Sort by Shape: Same Shape	5-6	144-145

137C Grade 1, Chapter 12, Cluster A

Name _____

Sort by Size

Circle the shape that is bigger than the others.

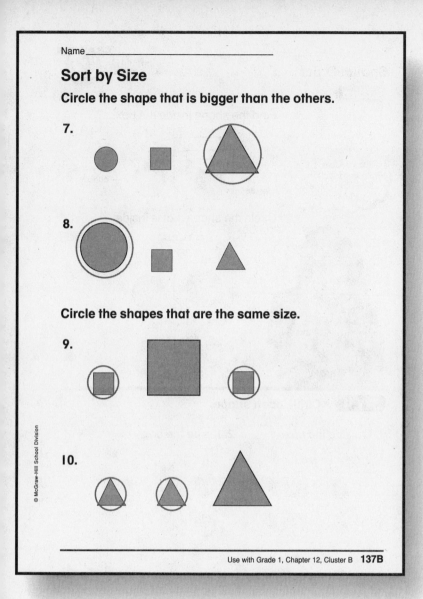

7.

8.

Circle the shapes that are the same size.

9.

10.

Use with Grade 1, Chapter 12, Cluster B **137B**

CLUSTER B PREREQUISITE SKILLS

The skills listed in this chart are those identified as major prerequisite skills for students' success in the lessons in Cluster B of the chapter. Each skill is covered by one or more assessment items as shown in the middle column. The right column provides the page numbers for the lessons in this book that reteach the Cluster B prerequisite skills.

Skill Name	Assessment Items	Lesson Pages
Sort by Size	7–10	146–147

CHAPTER 12 PRE-CHAPTER ASSESSMENT

Alternative Assessment Strategies

- Oral administration of the assessment is appropriate for younger students or those whose native language is not English. Read the skills title and directions one section at a time. Check students' understanding by asking them to tell you how they will do the first exercise in the group.
- For some skill types you may wish to use group administration. In this technique, a small group or pair of students complete the assessment together. Through their discussion, you will be able to decide if supplementary reteaching materials are needed.

Intervention Materials

If students are not successful with the prerequisite skills assessed on these pages, reteaching lessons have been created to help them make the transition into the chapter.

Item correlation charts showing the skills lessons suitable for reteaching the prerequisite skills are found beneath the reproductions of each page of the assessment.

Cluster B Challenge

Those students who demonstrate mastery of the skills on this page will not need to use the reteaching worksheets. Instead, these students can do the Cluster B Challenge found on pages 150–151.

Grade 1, Chapter 12, Cluster B **137D**

USING THE LESSON

Lesson Goal
- Recognize spatial order (outside, inside).

What the Student Needs to Know
- Use figure-ground perception.
- Recognize the meaning of "outside" and "inside."
- Read the words *outside* and *inside*.

Getting Started
Set up a cardboard box large enough for a student to step into. Call on a volunteer to stand beside the box. Ask:
- *Is [student's name] inside or outside the box?* (outside)
- Help the student step inside the box. Ask:
- *Is [student's name] inside or outside the box?* (inside)
- Call on another volunteer. Have one student stand in the box and the other stand beside it. Ask:
- *Who is inside the box? Who is outside the box?*

What Can I Do?
Read the question and the response. Then read and discuss the example. Say:
- *Put your finger on the shape that is inside the box. What shape is it?* (a square)
- *Put your finger on the shape that is outside the box. What shape is it?* (a circle)
- *Why is the square circled?* (The directions say to circle the shape that is inside the box.)

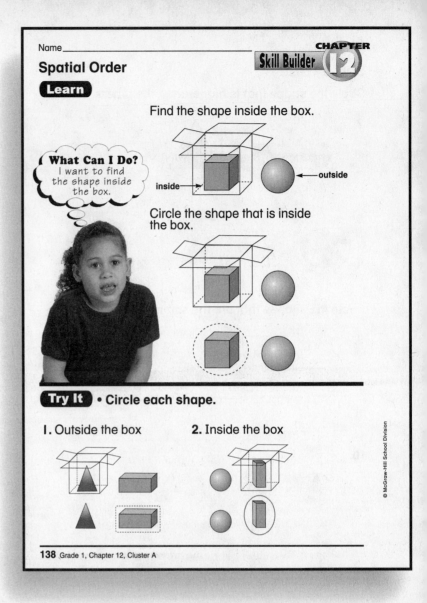

WHAT IF THE STUDENT CAN'T

Use Figure-Ground Perception
- Have the student place an attribute block or a pattern block inside a transparent box or container. Ask the student to describe the shape that is inside the container. Have the student repeat using different shapes.

Recognize the Meaning of "Outside" and "Inside"
- Give the student counters and a shoe box or other small box. Ask the student to place 3 counters inside the box; 2 counters outside the box; 5 counters inside the box 4 counters outside the box; and so on.
- Provide a paper bag and two toys; for example, a small doll and a toy truck. Ask the student to manipulate the toys so that the truck is inside the bag and the doll is outside; the doll is inside and the truck is outside; both are outside, and both are inside.

Name _____

Power Practice • Circle each shape.

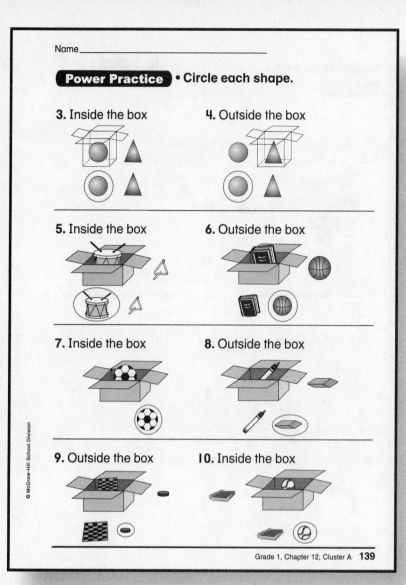

3. Inside the box
4. Outside the box
5. Inside the box
6. Outside the box
7. Inside the box
8. Outside the box
9. Outside the box
10. Inside the box

Grade 1, Chapter 12, Cluster A **139**

USING THE LESSON

Try It
- Read the directions aloud and explain what students are to do in each exercise.
- Do Exercise 1 with students.
- Ask: *What shape do you see outside the box?* (rectangle) Have students trace the circle around the rectangle.
- Have students complete Exercise 2 independently. Check their answers.

Power Practice
- Remind students that the directions may ask for the shape that is *inside* or *outside the box*. Students must read carefully to know which shape to choose.
- Have students complete the practice items. Review their answers.

WHAT IF THE STUDENT CAN'T

Read the Words *Outside* and *Inside*
- On index cards, write the words *outside* and *inside*. Have the student read the words. Place a toy inside a paper bag and have the student choose the correct card. Take the toy out and have the student choose the correct card. Continue until the student is clearly comfortable reading the words.

Complete the Power Practice
- Discuss each incorrect answer with the student. Have the student point to the shape that is called for by the directions. Then have the student circle the correct shape.

Grade 1, Chapter 12, Cluster A **139**

USING THE LESSON

Lesson Goal
- Use ordinal numbers to identify position.

What the Student Needs to Know
- Count to five.
- Match counting and ordinal numbers.

Getting Started
Give each student 5 counters. Ask:
- *How many counters do you have? Line them up and count them.* (5)
- *Which counter is first in line?* Remind students to move from left to right to determine position. *Which is last? Can you find the third counter in line?*
- Explain to students that in order to find the third counter, they should start at the left and count 3 counters.

What Can I Do?
- Read the question and the response. Then read and discuss the example. Ask:
- *Which animal is first?* (bird) *Which animal is third?* (rabbit)
- Ask students to point to the animals as you count aloud: *first, second, third, fourth, fifth.*
- Point out the similarities between some counting numbers and ordinals; for example, *third* begins with the 2 same letters as *three*, *fourth* has the same 4 letters as *four*, and *fifth* begins with the same 2 letters as *five*.
- Have students trace the x on the frog.

WHAT IF THE STUDENT CAN'T

Count to Five
- Provide the student with a pile of counters. Have the student count off five.
- Ask the student to fill in numbers on a blank 0–5 number line.

Match Counting and Ordinal Numbers
- Draw a row of five different shapes and number them 1–5. Have the student point to each shape from left to right and say its ordinal number: *first, second, third, fourth, fifth.*
- Make flash cards for 1–5 and *first-fifth*. Mix the cards up and have the student match each counting number to its ordinal number.

Name_____

Power Practice • Draw an X on the animal.

3. Draw an X on the first animal.

4. Draw an X on the second animal.

5. Draw an X on the fifth animal.

6. Draw an X on the fourth animal.

USING THE LESSON

Try It
- Read the directions aloud. Help students complete the first exercise.
- Ask: *Which animal are you to draw an X on?* (the third) *Count with me: first, second, third. Which animal is it?* (bird)
- Have students complete exercise 2 independently. Read students' answers.

Power Practice
- Depending on the reading ability of your class, you may wish to read all of the direction lines aloud and have students mark their answers one at a time.
- Have students complete the practice items. Then review each answer.

WHAT IF THE STUDENT CAN'T

Complete the Power Practice
- For each incorrect exercise, have the student identify the ordinal position of each animal in line. Then have the student draw an X on the appropriate animal.

USING THE LESSON

Lesson Goal
- Differentiate between shapes that roll and shapes that do not roll.

What the Student Needs to Know
- Relate two-dimensional drawings to three-dimensional figures.
- Recognize the meaning of "round."
- Identify corners.

Getting Started
Call on two volunteers. Give one a ball and one a cube-shaped block. Say:
- We are going to have a race. We will try to roll these objects to the finish line. Which shape do you think will go farther? (the ball)
- Mark a starting line on the floor with chalk. Have students "roll" their objects from the starting line.
- Ask: Which shape went farther? (the ball) Why? (It is round and can roll.)

What Can I Do?
- Read the question and the response. Then read and discuss the examples. Ask:
- Which shape is round, a spool or a block? (a spool)
- Which shape rolls, a spool or a block? (a spool)
- Which shape has corners, a spool or a block? (a block)
- Have students trace the cirdle aroundthe spool.

WHAT IF THE STUDENT CAN'T

Relate Two-Dimensional Drawings to Three-Dimensional Figures
- Give the student pairs of objects, one that rolls and one that does not. Examples might include an orange and a square cracker, a pencil and a book, a soda can and a milk carton. Have the student identify the object in each pair that rolls.
- Have the students cut out magazine pictures of objects that roll and objects that do not roll. Have the student sort them into two piles.

Recognize the Meaning of "Round"
- Ask the student to look around the classroom and find three things that are round.
- Display a cone, a cylinder, a sphere, a cube, and a triangular prism or pyramid. Ask the student to identify the figures that have round parts.

142 Grade 1, Chapter 12, Cluster A

USING THE LESSON

Try It
- Read the directions aloud and ask a volunteer to explain what students are to do.
- Complete exercise 1 with students. Ask: *What do the directions ask you to find?* (the shape that rolls) *Which shape is round and has no corners?* (the drum) *Which shape should you circle?* (the drum)
- Have students complete exercises 2-4 independently. Check their work.

Power Practice
- Point out how the directions change for items 11-14. Have students complete the practice exercises. Check students' answers.

WHAT IF THE STUDENT CAN'T

Identify Corners
- Provide the student with a cube and a rectangular prism. Have the student point to and count the corners of each shape.

Complete the Power Practice
- For each incorrect exercise, have the student identify the object that rolls and the object that does not roll. If possible, have the student use similar objects from the classroom to experiment with. Then have the student circle the oppropriate object.

USING THE LESSON

Lesson Goal
- Identify two-dimensional figures with the same shape.

What the Student Needs to Know
- Recognize two-dimensional shapes.
- Recognize attributes of shapes.
- Recognize the meaning of "same."

Getting Started
- Display two attribute blocks: a circle and a triangle.
- Ask: *How are these shapes different?* (One is round; one has corners.)
- Add a second triangle to the display.
- Ask: *Which two shapes are the same? Which shape is different?* (The triangles are the same; the circle is different.)

What Can I Do?
- Read the question and the response. Then read and discuss the example. Ask:
- *What four shapes do you see?* (circle, triangle, circle, square)
- *How is a circle different from a triangle? How is a circle different from a square?* (A circle is round; it has no corners.)
- *How is a triangle different from a square?* (A triangle has only 3 sides and 3 corners; a square has 4 sides and 4 corners.)

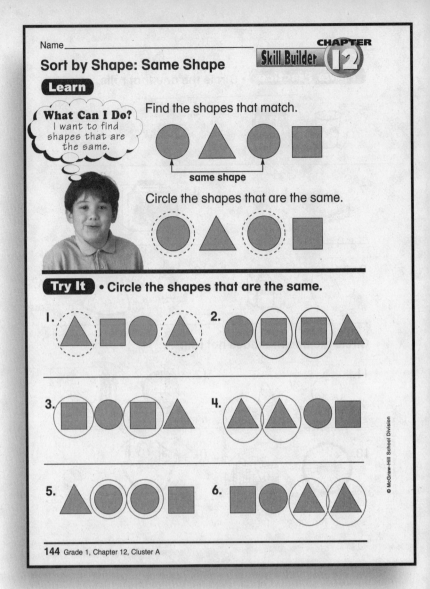

144 Grade 1, Chapter 12, Cluster A

Name_____

Power Practice • Circle the shapes that are the same.

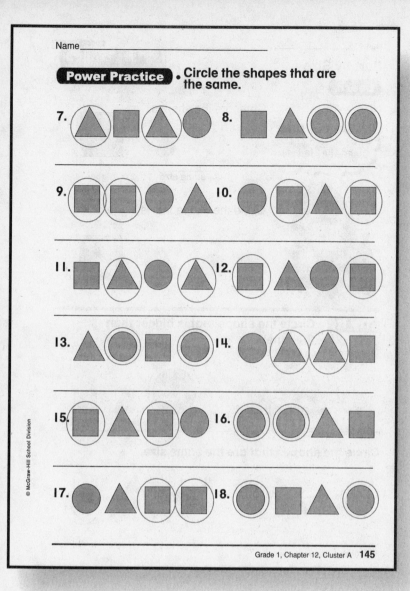

Grade 1, Chapter 12, Cluster A **145**

WHAT IF THE STUDENT CAN'T

Recognize the Meaning of "Same"

Distribute attribute blocks. Have the student find two blocks that are:
- the same color
- the same size
- the same shape

Complete the Power Practice
- Review any incorrect exercises with the student. Using attribute blocks, have the student model each exercise and identify the shapes that are the same. Then have the student circle the correct answers.

USING THE LESSON

Try It
- Read the directions aloud. Do the first exercise with students. Have students identify the shapes that are circled. (triangles)
- Ask: *Why are the triangles circled?* (They are the same shape.) *How can you tell they are the same?* (They both have 3 sides and 3 corners.)
- Have students complete the remaining exercises. Review their work.

Power Practice
- Remind students to think about roundness, corners, and sides as they complete this exercise.
- Read the directions. Have students complete the practice items. Then review each answer.

Learn with Partners & Parents

Tell students to go on a Shape Hunt in the classroom or at home to find objects that have these shapes:
- circle
- square
- triangle

Have students record the items they found and share their lists with the class.

Grade 1, Chapter 12, Cluster A **145**

USING THE LESSON

Lesson Goal
- Compare sizes of shapes.

What the Student Needs to Know
- Recognize two-dimensional shapes.
- Differentiate between big and small.
- Recognize the meaning of "same."

Getting Started
- Display the following attribute blocks circle and square.
- Ask: *How are these shapes different?* (One is round; one has sides and corners.)
- Add a larger square to the display.
- Ask: *Which two shapes are the same?* (the squares) *How are the squares different?* (One square is bigger than the other.)

What Can I Do?
- Read the question and the response. Then read and discuss the example. Ask:
- *What three shapes do you see?* (triangle, circle, square)
- *How are the triangle and circle the same?* (They are both small; they are the same size.)
- *How is the square different from the other two shapes?* (It is a different shape and a different size; it is bigger.)

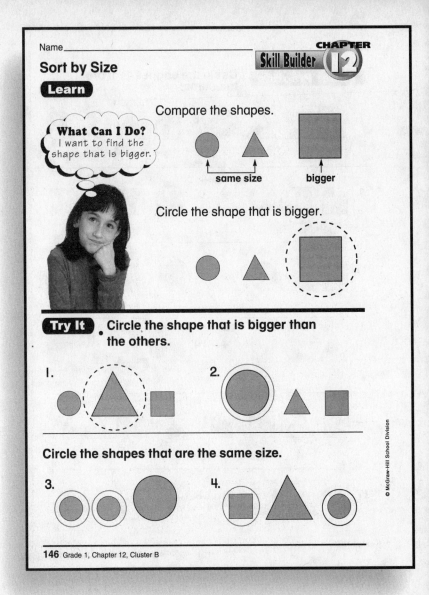

WHAT IF THE STUDENT CAN'T

Recognize Two-Dimensional Shapes
- Have the student skim pages 146–147 to find an exercise that shows only circles (3), an exercise that only shows squares (5 or 12), and an exercise that only shows triangles (11).

Differentiate Between Big and Small
- Distribute triangle attribute blocks in two sizes. Have the student sort them into groups of big and small triangles.
- Mix the blocks above and add circles in two sizes. Have the student sort all the blocks into piles of big and small blocks.

Name_____

Power Practice — Circle the shape that is bigger than the others.

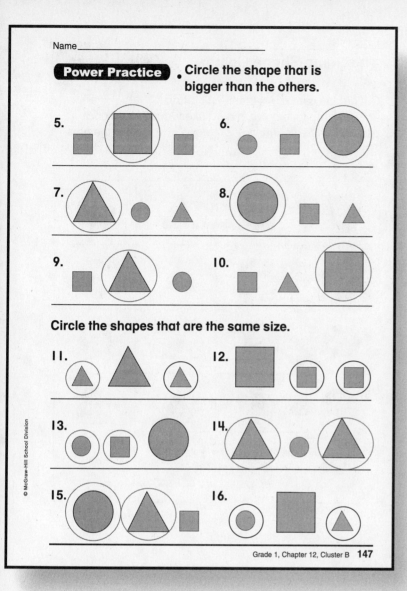

Circle the shapes that are the same size.

Grade 1, Chapter 12, Cluster B **147**

USING THE LESSON

Try It

- Read the two direction lines aloud. Have students identify the shape that is circled in exercise 1. (triangle)
- Ask: *Why is the triangle circled?* (It is bigger.) Have students trace the circle around the triangle.
- Have students complete the remaining exercises and share their answers with the group.

Power Practice

- Point out and read the two different direction lines before students begin. Have students complete the practice exercises. Then check their answers.

WHAT IF THE STUDENT CAN'T

Recognize the Meaning of "Same"

- Give the student a book from the classroom library and challenge the student to find another book that is the same size.
- Display a variety of attribute blocks and challenge the student to find two that are the same size, shape, and color.

Complete the Power Practice

- Review any incorrect exercises with the student. Using attribute blocks, have the student model each exercise, and then identify the shape that is bigger than the others or identify the shape that are the same size. Then have the student circle the correct answers.

Grade 1, Chapter 12, Cluster B **147**

CHALLENGE

Lesson Goal
- Match shapes to play a game.

Introducing the Challenge
- Give each group of students a pile of attribute blocks and have them sort the pile into triangles, circles, and squares.
- Distribute pages 148–149 to pairs of students. Have them cut apart the cards on page 148. Explain that they will use these shapes to play a game.

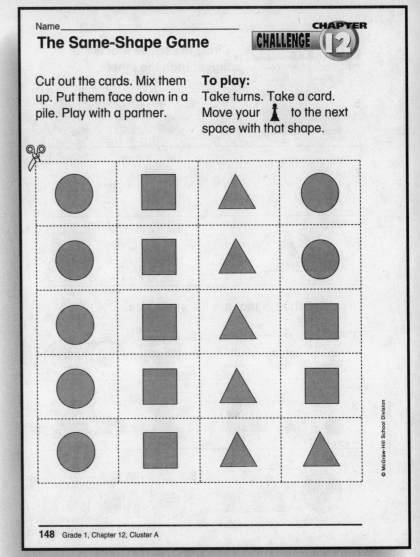

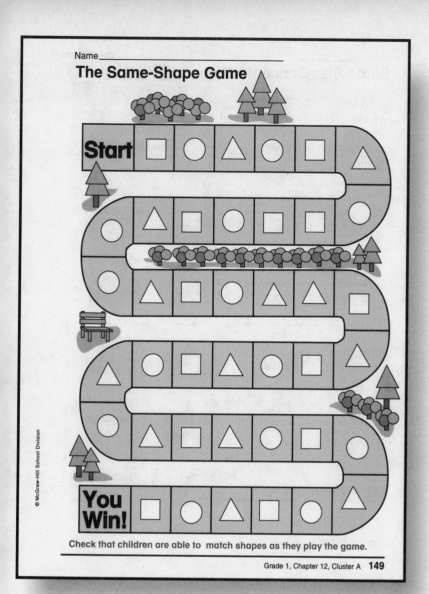

CHALLENGE

Using the Challenge

- Point out the "Start" and "You Win!" squares on the game board on page 149. Explain that the goal of the game is to reach "You Win!" by moving a marker along the path. Picking a shape card tells each player where to move.

- Give each student a marker. Have pairs mix the shape cards and place them face down in a pile. Say:

- *Take turns. Pick a card. Move your marker to the next game board square that shows that shape. The first player to reach the end wins.*

- After students have played the game once or twice, discuss how they found the shape on the game board that matched the shape card they picked.

CHALLENGE

Lesson Goal
- Match same-size shapes to play a game.

Introducing the Challenge
- Give each group of students a pile of attribute blocks in two sizes and have them sort the pile into small shapes and big shapes.
- Distribute page 150 to partners. Have them identify the shapes on the cards and tell whether they are big or small. Then have students cut out the shape cards.

Using the Challenge
- Have the pairs mix the cards and place them face down. Suggest that they place them in four rows of five cards. Read the directions aloud. Point out that a big circle and a big square can be a pair, and so can a small star and a small triangle. The only attribute they are matching is size.
- After students have played the game, they can play with this rule change: Match the shapes that are the same size and shape. Ask: *Which game was harder? Why?*

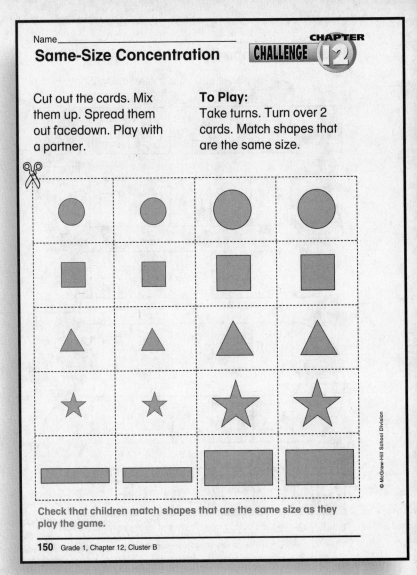

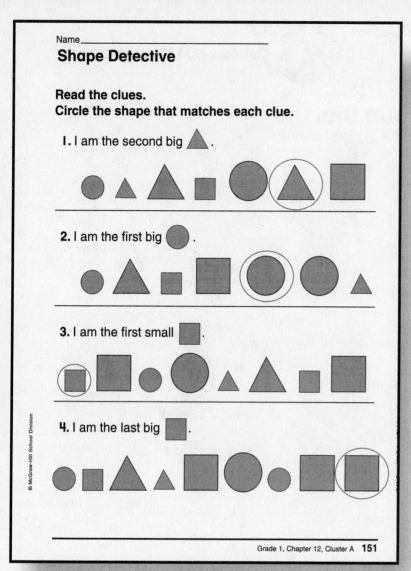

CHALLENGE

Lesson Goal
- Use logical reasoning to solve problems involving shape, size, and order.

Introducing the Challenge
- Set out two chairs. Have four students line up so that the first one is standing up, the second is sitting, the third is standing, and the fourth is sitting.
- Ask: *Who is the first standing person? Who is the second standing person? Who is the first sitting person? Who is the second sitting person?*

Using the Challenge
- Read the directions aloud. Have students who can read well complete the worksheet independently. For the others, you may wish to read each clue aloud.
- Point out that the clues require students to think about the order, the size, and the shape of the items pictured. If they don't consider all three attributes, they will not circle the correct shape. For example, if they don't consider order in exercise 1, they might circle the first big triangle. If they don't consider size in exercise 2, they might circle the first small circle. If they don't consider shape in exercise 2, they might circle the first big triangle.
- As a follow-up activity, students might like to make rows of attribute blocks and give clues to a partner. The partner has to find the block that matches the clues.

Grade 1, Chapter 12, Cluster B **151**

Name_____

Sort by Size

CHAPTER 13 — What Do I Need To Know?

Circle the ones in each group that are the same size.

1.

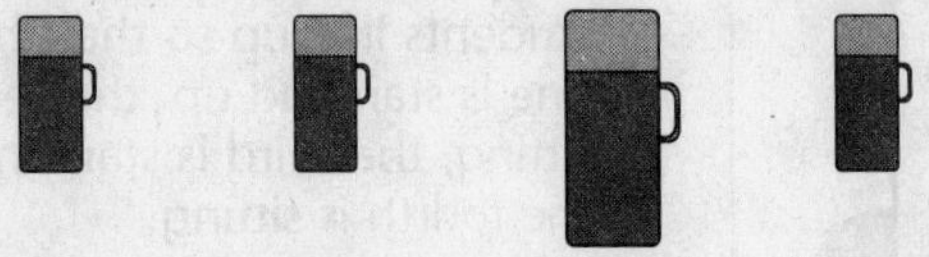

2.

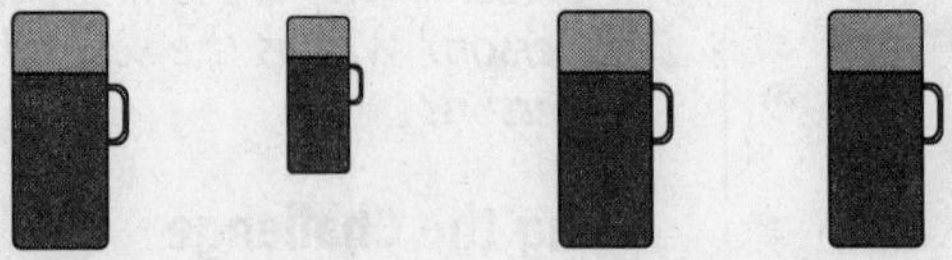

3.

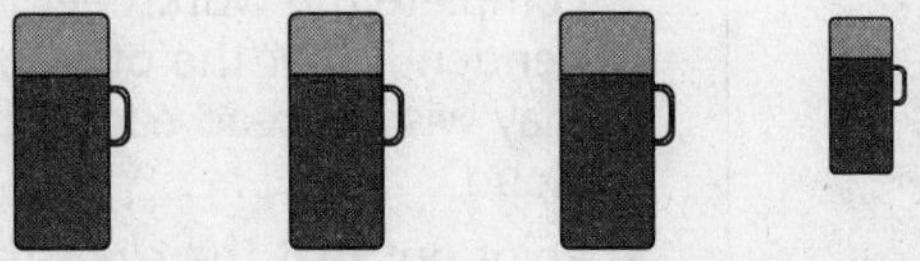

Equal Number

Circle the boxes that have the same number of paintbrushes.

4.

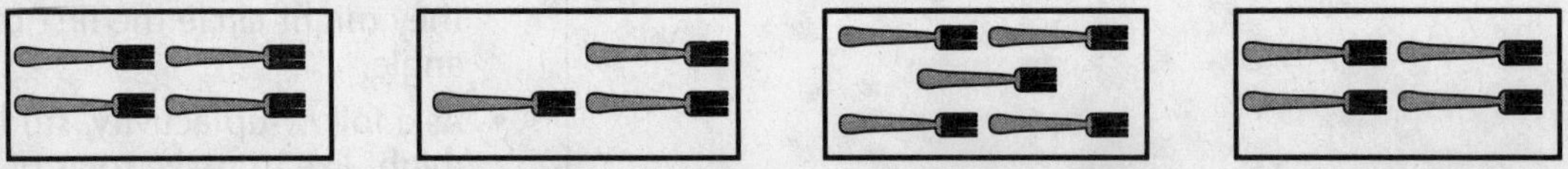

5.

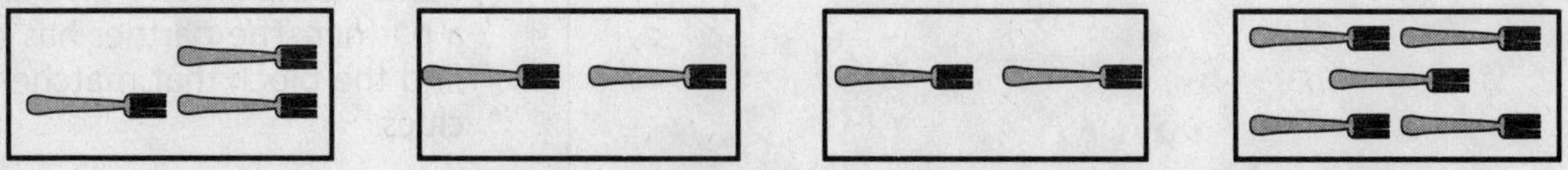

6.

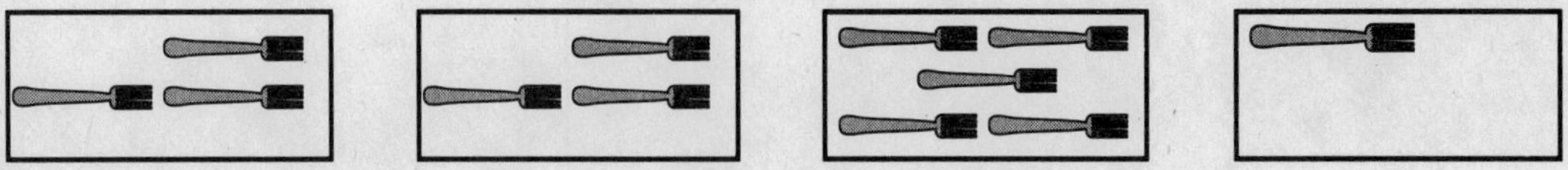

151A Use with Grade 1, Chapter 13, Cluster A

Name_____

Count to 10

Write how many.

7.

    _____

8.

    _____

9.

    _____

10.

    _____

CHAPTER 13 PRE-CHAPTER ASSESSMENT

Assessment Goal

This two-page assessment covers skills identified as necessary for success in Chapter 13 Fractions. The first page assesses the major prerequisite skills for Cluster A. The second page assesses the major prerequisite skills for Cluster B. When the Cluster A and Cluster B prerequisite skills overlap, the skill(s) will be covered in only one section.

Getting Started

- Allow students time to look over the two pages of the assessment. Point out the labels that identify the skills covered.
- Have students find math vocabulary terms used in the assessment. List vocabulary terms on the board as students identify them. If necessary, review the meanings of all essential math vocabulary.

Introducing the Assessment

- Explain to students that these pages will help you know if they are ready to start a new chapter in their math textbooks.
- Students who have transferred from another school may not have been introduced to some of these skills. Encourage students to do their best and assure them you will help them learn any needed skills.

Cluster A Challenge

Those students who demonstrate mastery of the skills on this page will not need to use the reteaching worksheets. Instead, these students can do the Cluster A Challenge found on page 158.

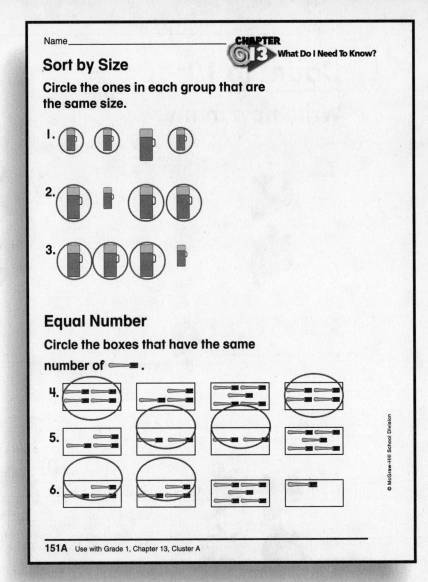

CLUSTER A PREREQUISITE SKILLS

The skills listed in this chart are those identified as major prerequisite skills for students' success in the lessons in Cluster A of the chapter. Each skill is covered by one or more assessment items as shown in the middle column. The right column provides the page numbers for the lessons in this book that reteach the Cluster A prerequisite skills.

Skill Name	Assessment Items	Lesson Pages
Sort by Size	1–3	152–153
Equal Number	4–6	154–155

151C Grade 1, Chapter 13, Cluster A

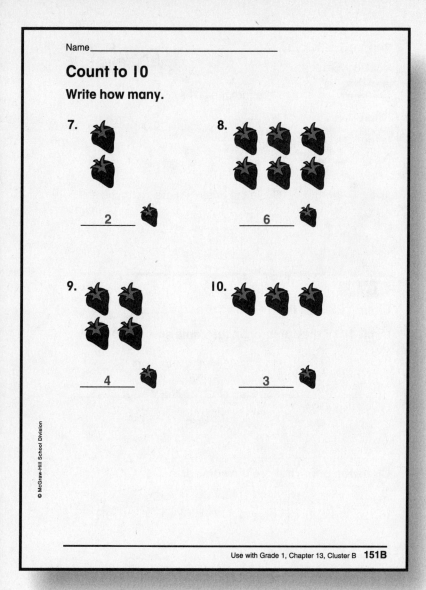

CHAPTER 13 PRE-CHAPTER ASSESSMENT

Alternative Assessment Strategies

- Oral administration of the assessment is appropriate for younger students or those whose native language is not English. Read the skills title and directions one section at a time. Check students' understanding by asking them to tell you how they will do the first exercise in the group.
- For some skill types you may wish to use group administration. In this technique, a small group or pair of students complete the assessment together. Through their discussion, you will be able to decide if supplementary reteaching materials are needed.

Intervention Materials

If students are not successful with the prerequisite skills assessed on these pages, reteaching lessons have been created to help them make the transition into the chapter.

Item correlation charts showing the skills lessons suitable for reteaching the prerequisite skills are found beneath the reproductions of each page of the assessment.

CLUSTER B PREREQUISITE SKILLS

The skills listed in this chart are those identified as major prerequisite skills for students' success in the lessons in Cluster B of the chapter. Each skill is covered by one or more assessment items as shown in the middle column. The right column provides the page numbers for the lessons in this book that reteach the Cluster B prerequisite skills

Skill Name	Assessment Items	Lesson Pages
Count to 10	7-10	156-157

Cluster B Challenge

Those students who demonstrate mastery of the skills on this page will not need to use the reteaching worksheets. Instead, these students can do the Cluster B Challenge found on page 159.

USING THE LESSON

Lesson Goal
- Sort objects by size.

What the Student Needs to Know
- Compare sizes of objects.
- Group objects according to their sizes.

Getting Started
- Use attribute blocks. Show students 3 large red squares. Ask:
- *Are all of these shapes the same?* (yes) *Are they the same color?* (yes, red) *Are they the same shape?* (yes, square) *Are they the same size?* (yes)
- Next, set out 2 large blue squares and 1 small blue square. Ask:
- *Are all of these shapes the same?* (no) *Why aren't they the same?* (One is small, and 2 are big.)

What Can I Do?
- Read the question and the response. Then read and discuss the example.
- Say: *Let's look at these 4 thermoses. Are they all the same? Why not?* (One is small, and 3 are big.)
- *If we circle the ones that are the same size, should we circle the small thermos?* (no) *Why not?* (It isn't the same size as the other 3 thermoses.)
- Have students trace the circles around the 3 large thermoses.

Try It
- Read the directions for exercises 1–2. Make sure students know what they are supposed to do. Work through the first exercise with students.
- Ask: *What do you see in the picture?* (4 thermoses) *Are all of them the same size?* (no) *Should we circle YES or NO?* (NO)
- Then have students complete exercise 2. Check their work.
- Read the directions for exercises 3–4. Make sure students know what to do. Work through exercise 3 with students.

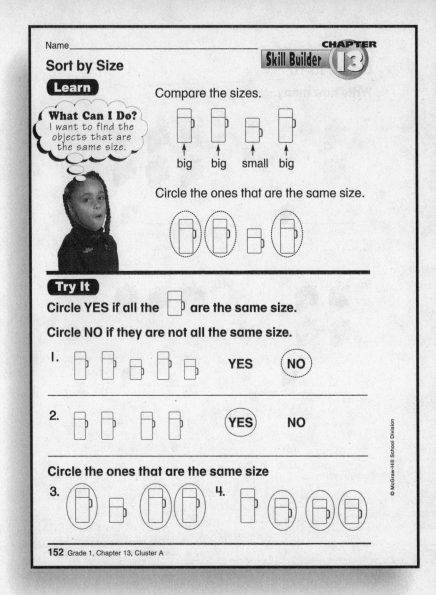

WHAT IF THE STUDENT CAN'T

Compare Sizes of Objects
- Show students pairs of objects that look the same but are different sizes (soup spoons and teaspoons; students' chairs and teachers' chairs; pairs of attribute blocks that are the same color and shape, but different sizes). For each pair, ask which object is bigger and which is smaller.
- From a collection of attribute blocks, have students identify pairs that differ in size only.
- Whenever appropriate, ask students questions about objects' sizes. For example, *Which is smaller, this safety pin or this pencil? Which is bigger, this chair or that one?*

Name_____

Power Practice — Circle YES if all the 🍶 are the same size. Circle NO if they are not all the same size.

5. 🍶 🍶 🍶 🍶   (YES)   NO

6. 🍶 🍶 🍶 🍶   YES   (NO)

7. 🍶 🍶 🍶 🍶   YES   (NO)

Circle the ones that are the same size.

8.   9.

10.   11.

Grade 1, Chapter 13, Cluster A **153**

USING THE LESSON

- **Ask:** *What do you see in the picture?* (4 thermoses, 3 big and 1 small) *Which ones are the same size?* (the 3 big ones) *Should we circle the little one?* (No, because it isn't the same size.)
- Have students complete exercises 3–4. Check their work.

Power Practice
- Read the two sets of directions with students. Then have students complete the practice exercises. Review students' answers.

Learn with Parents & Partners
- Have students play the following game with attribute blocks: Player A sets out several blocks. All of the blocks should be identical or they should differ in size only. For example, Player A might set out 6 large blue squares, *or* he or she might set out 4 large blue squares and 2 small blue squares.
- Player B first tells whether all of the blocks are the same size. If they are not, he or she groups them by size. After each turn, players switch roles.

WHAT IF THE STUDENT CAN'T

Group Objects According to Their Sizes
- Have students practice sorting attribute blocks into groups of large and small blocks that match in all other ways.
- Students might also practice sorting cut-out animal pictures from nature magazines into groups such as Small Animals, Medium-sized Animals, and Large Animals. Decide together what sizes each group should include. For example, anything smaller than a dog might be called "small."

Complete the Power Practice
- Discuss each incorrect answer. Have the student point to each object and say whether it is big or small. Then have the student correct his or her answer.

Grade 1, Chapter 13, Cluster A **153**

USING THE LESSON

Lesson Goal
- Match groups with the same number of objects.

What the Student Needs to Know
- Match objects in one-to-one correspondence.
- Identify the same number shown by different arrangements of objects.

Getting Started
- Use counters. Show students one group of 3, one group of 2, one group of 4, and a second group of 3. Say:
- *Let's count the counters in each group. Count the first group with me: 1-2-3. Now let's count the next group: 1-2. And the next one: 1-2-3-4. And the last one: 1-2-3.*
- *Do any of the groups have the same number of counters?* (Yes, there are two groups of 3.)
- Ask students who are wearing shirts with buttons to count the buttons. Have students with the same numbers of buttons stand together. Say, for example: *Everyone in this group has five buttons.*
- Alternatively, have students close their eyes and hold up 1, 2, 3, 4, or 5 fingers of one hand. Then have them open their eyes. Group the students so that those holding up the same number of fingers are together. Say, for example: *Everyone in this group is holding up 2 fingers.*

What Can I Do?
- Read the question and the response. Then read and discuss the example.
- Point to each group of sandwiches and have students count them with you. Point to each sandwich as you count: *Let's count the sandwiches in the first group: 1-2. Now let's count the next group: 1-2-3. The next one: 1-2. How many sandwiches does the last group have?* (1)

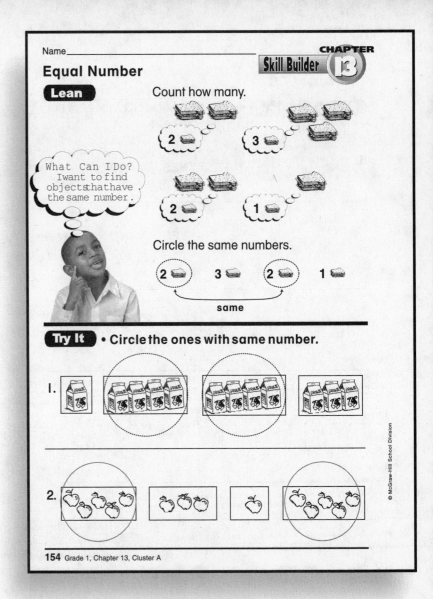

WHAT IF THE STUDENT CAN'T

Match Objects in One-to-One Correspondence
- Use counters of two different colors. Give the student the same number of each. Have him or her match the counters into pairs.
- Then have the student match objects such as 3 pencils and 3 sheets of paper, or 5 napkins and 5 plastic forks.
- Finally, have students match numbers to objects as they count them aloud.

Identify the Same Number Shown by Different Arrangement of Objects
- Use objects, counters, or connecting cubes. Name a number from 1–5. Then model that number in two different ways. For example, for 5, you might show a single counter above a row of 4 counters, and a row of 2 counters above a row of 3.
- Then have the student show two different ways to show other numbers.

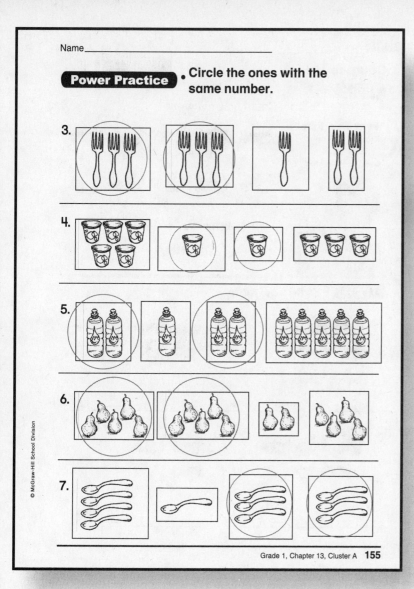

WHAT IF THE STUDENT CAN'T

Complete the Power Practice

- Discuss each incorrect answer. Have the student point to and count the objects in each group. Then have him or her tell which two groups have the same number. Finally, have the stuent circle the correct answers.

USING THE LESSON

- *Do any of the groups have the same number of sandwiches?* (Yes, two of the groups have 2 sandwiches each.)

Try It

- Read the directions and make sure students know what they are supposed to do.
- Work through the first exercise with students.
- Ask: *How many groups do you see in the picture?* (4 groups) *Do all of them have the same number of milk cartons?* (no) *How many milk cartons does each group have?* (The first group has 1, the second has 4, the third has 4, and the last group has 3.)
- *So which groups should we circle?* (the two groups with 4 cartons)
- *How can we check that we circled groups with the same number?* (Count the milk cartons.) *Let's count them: 1-2-3-4 — the first group we circled has 4. Let's count the other one: 1-2-3-4. They have the same number of cartons.*
- Then have students complete exercise 2. Check their answers.

Power Practice

- Read the directions with students. Say: *Look at exercise 3. How many forks does the first group have?* (3) *How many does the second group have?* (It also has 3.) *Do either of the other two groups have 3 forks?* (No, one has 1 fork and the other has 2.)
- *So which groups have the same number of forks?* (the first two) *Let's circle them.*
- Have students complete exercises 4–7. Check students' answers.

Grade 1, Chapter 13, Cluster A **155**

USING THE LESSON

Lesson Goal
- Count and write whole numbers through 10.

What the Student Needs to Know
- Match objects in one-to-one correspondence.
- Read, write, and count numbers to 10.
- Identify the same number shown by different arrangements of objects.

Getting Started
- Use a stack of 10 books. Ask students to count them with you. Pick up and restack each book as you count it. Say:
- *Let's count these books: 1-2-3-4-5-6-7-8-9-10.* Write 10 on the chalkboard.
- Suggest that students count something that has 10 or fewer items. For example, count pillows in a reading center or chairs around a table. Say: *Let's count these _____.* Then count them with students. A volunteer might place a hand on each item as everyone counts. Have another volunteer write the number on the chalkboard.

What Can I Do?
- Read the question and the response. Then read and discuss the example.
- Say: *Let's count these picnic baskets. Put your finger on each basket as you count it. 1-2-3-4-5-6-7-8-9-10.*
- *How many baskets are there?* (10)
- *Now write the number of baskets — 10.* Have students trace the number.

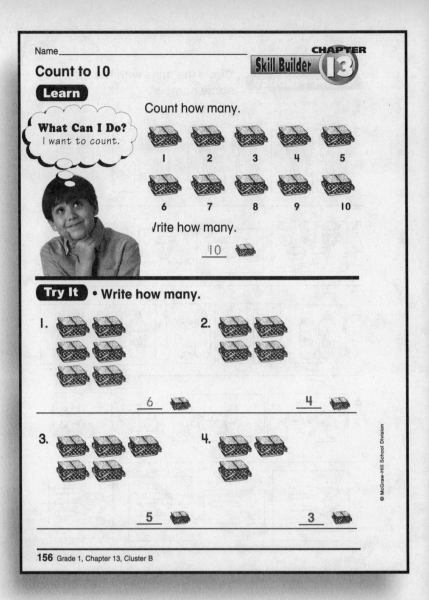

WHAT IF THE STUDENT CAN'T

Match Objects in One-to-One Correspondence
- Using counters of two different colors, give the student the same number of each (up to 10 of each color). Have the student match the counters into pairs.
- Then have the student match objects such as 7 cups and 7 napkins, or pictures such as 10 squirrels and 10 trees.
- Finally, have the student match numbers to objects as they count them aloud.

Read, Write, and Count Numbers to 10
- Set out 10 counters. Model counting them, and then have the student count. Next, change the number of counters and have the student count them. Continue using different numbers of counters.
- Have the student match a given number of objects with the appropriate number card. Repeat with several more groups (from 1–10 objects per group).

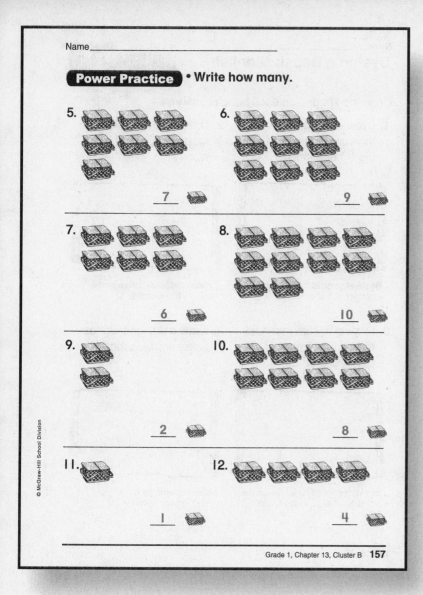

USING THE LESSON

Try It

- Read the directions and make sure students know what they are supposed to do.
- Work through the first exercise with students.
- Say: *Let's count the baskets together: 1-2-3-4-5-6. What should we write on the line?* (6)
- Then have students complete exercises 2–4. Check their work.
- For students who need more support with writing numbers, suggest that they look back at the numbers at the top of the page.

Power Practice

- Read the directions with students. Then have students complete the practice exercises. Review answers with students.
- Have volunteers choose a few exercises and demonstrate counting the picnic baskets. Volunteers can write the number of baskets on the chalkboard.

WHAT IF THE STUDENT CAN'T

Identify the Same Number Shown by Different Arrangement of Objects

- Use counters or connecting cubes. Name a number from 1–10. Model showing that number in two or more different ways. For example, for 10 you might show: a single cube and three 3-cube trains; five 2-cube trains; or two 5-cube trains.
- Then have students show two different ways to show other numbers.

Complete the Power Practice

- Discuss each incorrect answer. Have the student point to and count the baskets in the exercise. Then have him or her write the correct number.
- If the student has difficulty counting the pictured objects, have him or her count counters or connecting cubes grouped in the same arrangements.

Grade 1, Chapter 13, Cluster B **157**

CHALLENGE

Lesson Goal
- Use the concepts of big/small, more/fewer, and various shapes to create designs.

Introducing the Challenge
- Tell students they are going to decorate the beach blankets shown on page 158.
- Explain that for each design they make, students will need to follow directions very carefully.
- Work with students to be sure they understand the concepts of big/small and more/fewer. Have volunteers come to the chalkboard and demonstrate drawing squares, circles, and triangles.

Using the Challenge
- Read the general directions aloud. Next, read aloud the directions for exercise 1. Draw a large rectangle on the board and ask students to suggest ways to decorate a beach blanket with the same number of big and small circles.
- For example, you might draw one row of 6 big circles along the top of the rectangle, a row of 6 small circles beneath that, then repeat the pattern.
- Using students' suggestions, draw 2–3 different designs. After completing each one, have a volunteer count the circles to make sure there are the same number of big and small ones.
- Have students create their own designs for exercise 1. Check their work to be sure they followed directions.
- Read aloud the directions for exercise 2. Be sure students understand what they are supposed to do. Repeat for exercises 3–4.
- You may wish to display students' work in a bulletin board display. Students may wish to copy their best designs on art paper using colored markers.

Name_____

Design a Beach Blanket

Draw the shapes on each beach blanket.

1. Draw a design that has the same number of big ○ and small ○.

2. Draw a design that has more big △ than small △.

Blanket should have an equal number of big and small ○.

Blanket should have more big △ than small △.

3. Draw a design that has the same number of big △ and small □.

4. Draw a design that has fewer small △ than big ○.

Blanket should have an equal number of big △ and small □.

Blanket should have more big ○ than small △.

158 Grade 1, Chapter 13, Cluster A

Name_____

Domino Fun

Cut out the dominoes. Mix them up. Put them facedown in a pile. Play with a partner.

To play:
Take turns. Turn over 2 🁣 to start. Turn over 1 🁣 from the pile. Match the dots at one end to one of your 🁣. If you have no match, take a 🁣.

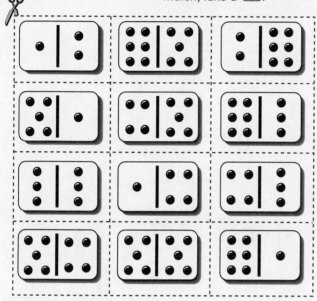

Check that children match the dominoes correctly.

CHALLENGE

Lesson Goal
- Match dot patterns that show numbers from 1–6.

Introducing the Challenge
- Tell students they are going to play a game called *Dominoes*. To play the game they will need to match dot patterns that show numbers from 1–6.
- In one column on the chalkboard, copy each of the dot patterns shown on the dominoes. As you point to each pattern, have students call out the number it represents.
- Next, draw the same dot patterns on the board in a second column opposite the first one. Draw the patterns in scrambled order. Have volunteers come to the board and draw lines to match the patterns in the first column with those in the second column.

Using the Challenge
- Read the directions aloud. You may wish to cut out a set of dominoes yourself and demonstrate how to play the game before students form pairs and play on their own.
- Drawing a domino from the pile, each player tries to match one of his or her dominoes to that domino. Players alternate drawing dominoes and matching.

CHAPTER 14 — What Do I Need To Know?

Add Sums to 20

Add. Write each sum.

1.  🌼 🌼 🌼 🌼 🌼 🌼 🌼
   🌼 🌼 🌼 🌼 🌼 🌼 🌼

   7 + 7 = _____ 🌼

2. 5 + 6 = _____

3.   7
    + 3
   ─────

Count to 100

4. Write the number of tens and ones.
   Write the number.

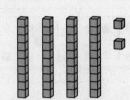

   _____ tens _____ ones = _____

Names for Numbers

Circle the models that show another way to name the number.

5. 12 ones

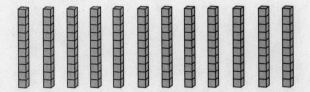

6. 3 tens 6 ones

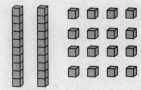   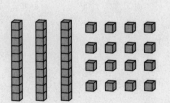

Name_____

Subtract from 20

Subtract. Write each difference.

7.  🌼 🌼 🌼 🌼 🌼
   🌼 🌼 🌼 🌼 🌼
   🌼 🌼

   12 − 3 = _____ 🌼

8. 16 − 8 = _____

9.   15
   − 6
   ─────

Subtract. Write each difference.

10. 11 − 8 = _____

    11 − 3 = _____

Use with Grade 1, Chapter 14, Cluster B   **159B**

CHAPTER 14 PRE-CHAPTER ASSESSMENT

Assessment Goal

This two-page assessment covers skills identified as necessary for success in Chapter 14 Add and Subtract 2-Digit Numbers. The first page assesses the major prerequisite skills for Cluster A. The second page assesses the major prerequisite skills for Cluster B. When the Cluster A and Cluster B prerequisite skills overlap, the skill(s) will be covered in only one section.

Getting Started

- Allow students time to look over the two pages of the assessment. Point out the labels that identify the skills covered.
- Have students find math vocabulary terms used in the assessment. List vocabulary terms on the board as students identify them. If necessary, review the meanings of all essential math vocabulary.

Introducing the Assessment

- Explain to students that these pages will help you know if they are ready to start a new chapter in their math textbooks.
- Students who have transferred from another school may not have been introduced to some of these skills. Encourage students to do their best and assure them you will help them learn any needed skills.

Cluster A Challenge

Those students who demonstrate mastery of the skills on this page will not need to use the reteaching worksheets. Instead, these students can do the Cluster A Challenge found on page 172.

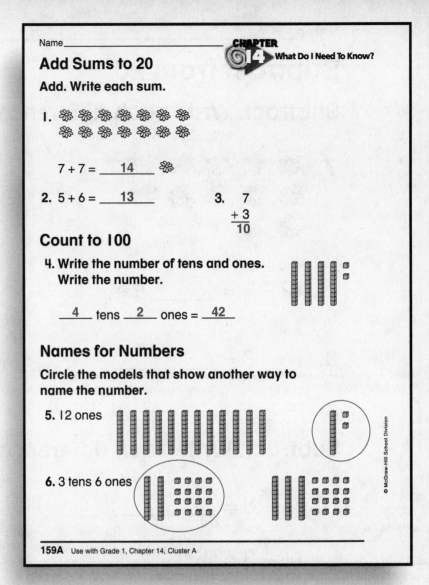

CLUSTER A PREREQUISITE SKILLS

The skills listed in this chart are those identified as major prerequisite skills for students' success in the lessons in Cluster A of the chapter. Each skill is covered by one or more assessment items as shown in the middle column. The right column provides the page numbers for the lessons in this book that reteach the Cluster A prerequisite skills.

Skill Name	Assessment Items	Lesson Pages
Add Sums to 20	1-3	160-163
Count to 100	4	164-165
Names for Numbers	5-6	166-167

Name_____

Subtract from 20

Subtract. Write each difference.

7. ❀❀❀❀❀
   ❀❀❀❀
   ❀❀

   $12 - 3 = \underline{9}$ ❀

8. $16 - 8 = \underline{8}$

9. $\begin{array}{r} 15 \\ -6 \\ \hline 9 \end{array}$

Subtract. Write each difference.

10. $11 - 8 = \underline{3}$

    $11 - 3 = \underline{8}$

Use with Grade 1, Chapter 14, Cluster B **159B**

CLUSTER B PREREQUISITE SKILLS

The skills listed in this chart are those identified as major prerequisite skills for students' success in the lessons in Cluster B of the chapter. Each skill is covered by one or more assessment items as shown in the middle column. The right column provides the page numbers for the lessons in this book that reteach the Cluster B prerequisite skills

Skill Name	Assessment Items	Lesson Pages
Subtract from 20	7-10	168-171

CHAPTER 14 PRE-CHAPTER ASSESSMENT

Alternative Assessment Strategies

- Oral administration of the assessment is appropriate for younger students or those whose native language is not English. Read the skills title and directions one section at a time. Check students' understanding by asking them to tell you how they will do the first exercise in the group.

- For some skill types you may wish to use group administration. In this technique, a small group or pair of students complete the assessment together. Through their discussion, you will be able to decide if supplementary reteaching materials are needed.

Intervention Materials

If students are not successful with the prerequisite skills assessed on these pages, reteaching lessons have been created to help them make the transition into the chapter.

Item correlation charts showing the skills lessons suitable for reteaching the prerequisite skills are found beneath the reproductions of each page of the assessment.

Cluster B Challenge

Those students who demonstrate mastery of the skills on this page will not need to use the reteaching worksheets. Instead, these students can do the Cluster B Challenge found on page 173.

Grade 1, Chapter 14, Cluster B **159D**

USING THE LESSON

Lesson Goal
- Add facts to 20.

What the Student Needs to Know
- Recognize the meaning of addition.
- Recognize plus and equal signs.
- Add horizontally or vertically.

Getting Started
- On a felt board, show a group of 3 objects and a group of 5 objects. Have students give the number sentence and solve for the sum. (3 + 5 = 8)
- Repeat with 5 and 7 objects, 8 and 2 objects, and 6 and 7 objects. (5 + 7 = 12, 8 + 2 = 10, 6 + 7 = 13)

What Can I Do?
- Read the question and the response. Then read and discuss the example. Ask:
- *How many flowers are in the first pot?* (4) *How many flowers are in the second pot?* (6) *How can you find how many flowers there are in all?* (Put them together; add.)
- Have students trace the sum.

Try It
- Ask: *How is the addition in exercise 1 different from the addition in exercise 2?* (One is a number sentence written in a row; the other is written in a column.)
- Do exercise 1 as a class. Say:
- *How many flowers are in the first pot?* (3) *How many flowers are in the second?* (5) *How many flowers are there in all?* (8) Have students trace the sum.

Power Practice
- Have students complete the practice. Review each answer.
- Ask students to tell how solving exercise 3 helps them solve exercise 4. (10 + 5 will be 1 more than 9 + 5.)

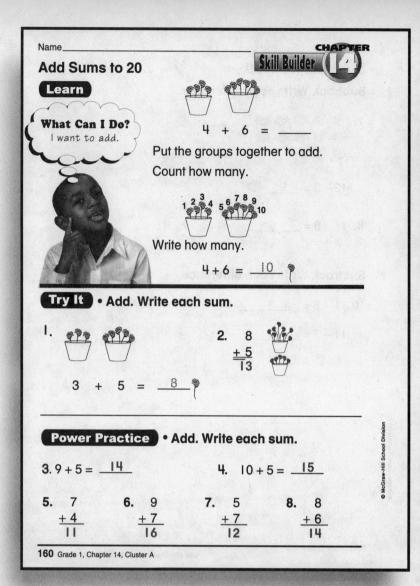

WHAT IF THE STUDENT CAN'T

Recognize the Meaning of Addition
- Have the student use the felt board and felt shapes to show these facts: 1 + 3, 4 + 5, 5 + 2.
- Ask the student to imagine that he or she is at a supermarket. Ask: *When might you need to add at the store?* (to find the amount of money you need to pay for food; to add the weight of two different foods, and so on)

Recognize Plus and Equal Signs
- Write number sentences with circles in place of the plus and equal signs. For example: 5 [circle] 4 [circle] 9. Have students supply the signs and read the sentences aloud.

Add Horizontally or Vertically
- Ask the student to rewrite exercises 3–4 as vertical addition problems and exercises 5–8 as number sentences. Discuss how this affects the sum. (It doesn't.)

Complete the Power Practice
- Discuss each incorrect answer. Have the student use counters to model any fact he or she missed.

Name_____

Add Sums to 20: Counting On

Skill Builder CHAPTER 14

Learn

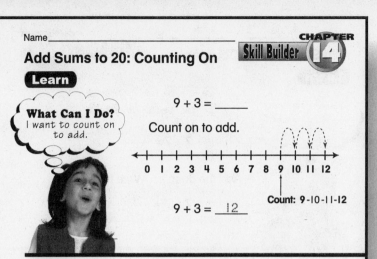

What Can I Do? I want to count on to add.

9 + 3 = ____

Count on to add.

0 1 2 3 4 5 6 7 8 9 10 11 12

Count: 9-10-11-12

9 + 3 = __12__

Try It
• Count on to add. Write each sum.

1. 8 + 3 = __11__

2. 10
   + 2

     12

3. 9
   + 1

    10

Power Practice
• Count on to add. Write each sum.

4. 7 + 2 = __9__

5. 6 + 3 = __9__

6. 5   7. 8   8. 7   9. 10
   + 1   + 2   + 3   + 1
   ---   ---   ---   ---
     6   10   10   11

Grade 1, Chapter 14, Cluster A **161**

WHAT IF THE STUDENT CAN'T

Count to Twelve
- Mix number cards 0-12 and have the student line them up in order.
- Place a number of counters between 1 and 12 on the table. Ask the student to count them.

Read a Number Line
- Draw a 0-12 number line on the chalkboard. Omit some of the numbers and have the student replace them.

Add One
- Use a 0–12 number line. Ask the student to find the number 10 and count on one.

Ask: *What number is that?* (11) Repeat with other numbers greater than 10. Have the student write the number sentence each time (10 + 1 = 11, and so on.)

Complete the Power Practice
- Discuss each incorrect answer. Have the student use the number line to find the sum. Then have the student write the correct answer.

USING THE LESSON

Lesson Goal
- Count on to add.

What the Student Needs to Know
- Count to twelve.
- Read a number line.
- Add one.

Getting Started
- Have students count aloud from 0 to 12. Then have them count to 12 beginning with 5, beginning with 3, and beginning with 8.

What Can I Do?
- Read the question and the response. Then read and discuss the example. Ask:
- *When you add 9 + 3, where do you start on the number line?* (at 9)
- *How many steps do you count on to add 3?* (3 steps)
- *Suppose you add 2 + 8. Where do you start?* (2) *How many steps do you count on?* (8)
- Have students trace the sum at the end of the example.

Try It
- Students may use the number line to find the sums. Give them the following steps to follow as you review exercise 1:
- *Find the first number on the number line. Put your finger there.*
- *To add 3, count on 3 spaces. Put your finger there. What is the sum?* (11) Have students trace the number.
- Have students complete exercises 2 and 3. Check their answers.

Power Practice
- Tell students that they may use the number line if they wish.
- Have students complete the practice items. Then review each answer.

Grade 1, Chapter 14, Cluster A **161**

USING THE LESSON

Lesson Goal
- Add doubles with sums to 20.

What the Student Needs to Know
- Match one-to-one.
- Count to 20.
- Recognize plus and equal signs.

Getting Started
- Give each pair of students 10 counters. Have the students arrange 4 of the counters in 2 rows of 2 counters.
- On the chalkboard, write $2 + 2 = 4$.
- Have the students add 1 counter to each row. Write $3 + 3 = 6$.
- Repeat twice, writing $4 + 4 = 8$ and $5 + 5 = 10$.
- Discuss why the facts you wrote are called "doubles." (You double the number being added.)

What Can I Do?
- Read the question and the response. Then read and discuss the example. Ask:
- *How can drawing a picture help you solve doubles?* (Use the picture to count and find the sum.)
- *How would knowing 7 + 7 help you know 7 + 8?* (Because 8 is 1 more than 7, the answer will be 1 more than 14, or 15.)

Try It
- Explain to students that memorizing the sums of doubles will help them with all kinds of related addition facts.
- Do exercise 1 with students. Have them complete exercise 2 independently.

Power Practice
- Read the directions. Then have students complete the practice items. Review each answer.
- Challenge students to find the sum on the page that is greatest ($10 + 10 = 20$) and least ($2 + 2 = 4$).

Name _____

Add Sums to 20: Doubles

Skill Builder — CHAPTER 14

Learn

What Can I Do? I want to add doubles.

$7 + 7 = $ _____

Add the doubles. Write the sum.

$7 + 7 = \underline{14}$

Try It • Add the doubles. Write the sum.

1.

   $5 + 5 = \underline{10}$

2. $\begin{array}{r} 3 \\ +3 \\ \hline 6 \end{array}$

Power Practice • Add the doubles. Write the sum.

3. $9 + 9 = \underline{18}$

4. $6 + 6 = \underline{12}$

5. $\begin{array}{r} 8 \\ +8 \\ \hline 16 \end{array}$

6. $\begin{array}{r} 4 \\ +4 \\ \hline 8 \end{array}$

7. $\begin{array}{r} 10 \\ +10 \\ \hline 20 \end{array}$

8. $\begin{array}{r} 2 \\ +2 \\ \hline 4 \end{array}$

WHAT IF THE STUDENT CAN'T

Match One-to-One
- On the chalkboard or on paper, draw parallel rows of 6 dots and 6 dots, 7 dots and 8 dots, 4 dots and 4 dots, 6 dots and 5 dots, and 8 dots and 8 dots. Have students identify the doubles and find the sums.
- Lay out a row of 1-9 counters. Have the student make an equivalent row, write the double, and find the sum.

Count to 20
- Give the student a large pile of counters or pennies and have him or her count out 20.
- Have the student count aloud to 20 from 1, from 4, and from 10.

Recognize Plus and Equal Signs
- Write a variety of addition number sentences on the board and have the student read them aloud.

Complete the Power Practice
- Discuss each incorrect answer. Have the student use parallel rows of counters to model any fact he or she missed. Then have the student write the correct sum.

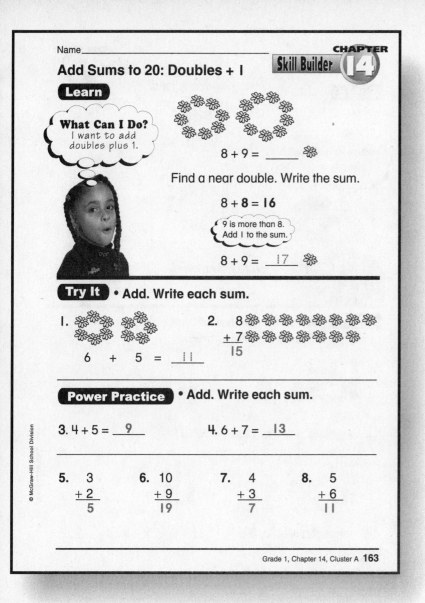

USING THE LESSON

Lesson Goal
- Add doubles plus 1.

What the Student Needs to Know
- Add doubles.
- Add one.

Getting Started
- Give each pair of students 7 counters. Have the students arrange 4 counters in 2 rows of 2 counters.
- On the board, write 2 + 2 = 4.
- Have students add 1 counter to one of the rows. Write 3 + 2 = 5.
- Repeat with 3 + 3 = 6 and 4 + 3 = 7. Ask: *How does knowing the sums of doubles help you add other facts?* (If one of the numbers being added is 1 more than a double, the sum is 1 more than the sum of the doubles.)

What Can I Do?
- Read the question and the response. Then read and discuss the example. Ask:
- *Why is the sum of 8 and 9 one more than the sum of 8 and 8?* (9 is 1 more than 8, so the sum of 8 + 9 = 8 + 8 + 1.)
- *How would knowing 5 + 5 help you know 5 + 6?* (Because 6 is 1 more than 5, the answer will be 1 more than 10, or 11.)

Try It
- Have students identify the doubles they can use to help them find the sums in exercises 1 and 2. (5 + 5, 7 + 7)
- Do exercise 1 with the class. Have them trace the sum. Have them complete exercise 2 independently.

Power Practice
- Read the directions. Then have students complete the practice items. Review each answer.
- Ask students to identify the doubles that helped them solve exercises 3–8. (4 + 4, 6 + 6, 2 + 2, 9 + 9, 3 + 3, 5 + 5)

Grade 1, Chapter 14, Cluster A **163**

USING THE LESSON

Lesson Goal
- Write 2-digit numbers.

What the Student Needs to Know
- Count ten items.
- Differentiate tens models from ones models.

Getting Started
- Display a tens model and a ones model. Ask:
- *Which model stands for 1 one? Which model stands for 1 ten? What number do they make when we put them together?* (11) So, 1 ten and 1 one equals 11.
- Display 2 tens models and 2 ones models. Ask:
- *Which model stands for 2 ones? Which model stands for 2 tens? What number do they make when we put them together?* (22) So, 2 tens and 2 ones equals 22.

What Can I Do?
- Read the question and the response. Then read and discuss the example. Ask:
- *When you write a number, which is on the left, the tens or the ones?* (the tens)
- *How many tens are being shown by the models?* (5) Have students trace the number of tens.
- *How many ones are shown by the models?* (3) Have the students trace that number. Say: Trace the number for 5 tens 3 ones. Have students trace 53.

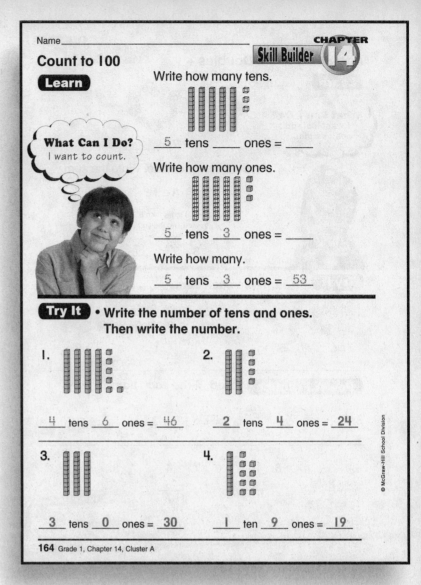

WHAT IF THE STUDENT CAN'T

Count Ten Items
- Place ten counters in a line. Have the student count them. Then have the student count ten connecting cubes and make a 10-cube train.
- Give the student a pencil and paper and ask him or her to draw a group of 6 dots, a group of 8 dots, and a group of 10 dots.

Differentiate Tens Models from Ones Models
- Give the student 9 ones models and have him or her count them by ones. Then give the student 9 tens models. Help him or her count them by tens.
- Ask the student to use tens and ones models to show the following numbers: 12 and 21, 34 and 43, 65 and 56, and so on.

Name_____

Power Practice • Write the numbers of tens and ones. Then write the number.

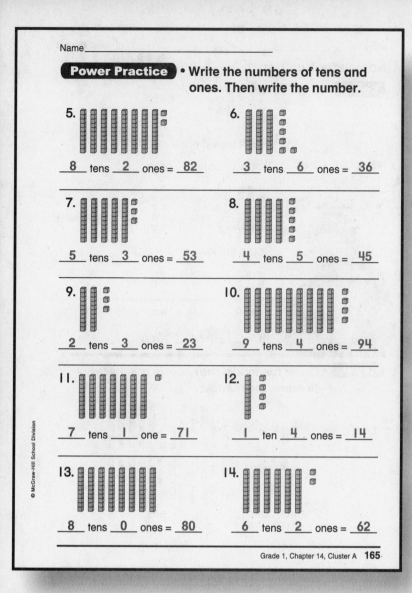

5. _8_ tens _2_ ones = _82_
6. _3_ tens _6_ ones = _36_
7. _5_ tens _3_ ones = _53_
8. _4_ tens _5_ ones = _45_
9. _2_ tens _3_ ones = _23_
10. _9_ tens _4_ ones = _94_
11. _7_ tens _1_ one = _71_
12. _1_ ten _4_ ones = _14_
13. _8_ tens _0_ ones = _80_
14. _6_ tens _2_ ones = _62_

Grade 1, Chapter 14, Cluster A **165**

USING THE LESSON

Try It

- Work through exercise 1 with the students. Say: *How many tens do you count?* (4) *Write 4 in the first space. How many ones do you count?* (6) *Write 6 in the next space. What number is 4 tens 6 ones?* (46) *Write that number.*
- Students who are having trouble might circle the tens models and then circle the ones models before writing the number.
- Have students complete exercises 2–4. Check students' answers.

Power Practice

- Read the directions. Remind students that each problem requires three answers: the number of tens, the number of ones, and the number.
- Have students complete the practice items. Discuss their answers.

WHAT IF THE STUDENT CAN'T

Complete the Power Practice

- Discuss each incorrect answer. Look for common errors such as reversed place value or incorrect counting.
- Have the student use tens and ones models to show each number. Then have the student write each corret answer.

Grade 1, Chapter 14, Cluster A **165**

USING THE LESSON

Lesson Goal
- Rename tens and ones as ones; rename ones as tens and ones.

What the Student Needs to Know
- Identify place value.
- Count ten items.
- Equate 1 ten to 10 ones.

Getting Started
- Distribute tens and ones models to small groups of students. Have students pick one tens model and show the number of ones equal to 1 ten. (10 ones) Continue with 1 ten 1 ones (11 ones), 1 ten 2 ones (12 ones), 1 ten 3 ones (13 ones), and so on.
- Ask: *How many ones are the same as 1 ten 8 ones?* (18 ones) *How many of tens and ones are the same as 19 ones?* (1 ten 9 ones)

What Can I Do?
- Read the question and the response. Then read and discuss the examples. Ask:
- *How many ones are there in 4 tens and 8 ones?* (48 ones) *How can you make 4 tens 8 ones into 3 tens 18 ones?* (Trade 1 ten for 10 ones.)
- *Suppose you had 3 tens 6 ones. What would you have if you traded 1 ten for 10 ones?* (2 tens 16 ones) *When you trade 1 ten for 10 ones, the ones go up by 10, and the tens go down by 1 ten.*

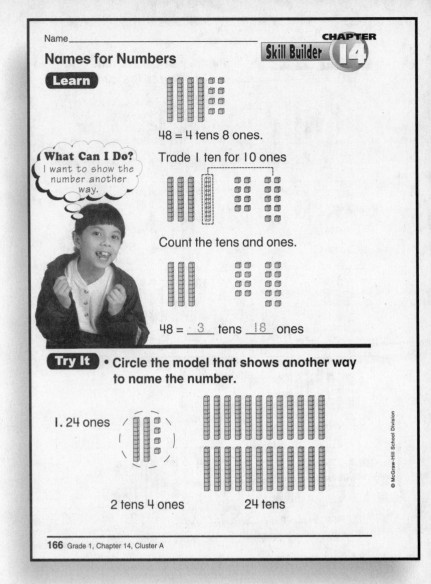

WHAT IF THE STUDENT CAN'T

Identify Place Value
- Give students a hundred chart. Have them choose five numbers at random and tell the number of tens and ones in each number. Then have them express that as a number of ones. For example, 24 = 2 tens 4 ones = 24 ones.

Count Ten Items
- Give pairs of students a pile of play pennies. Have them count the pennies and make stacks of ten. If you like, you may have them "cash in" the stacks of pennies for dimes.

Equate 1 Ten to 10 Ones
- Give pairs of students tens and ones models. Have them take turns counting out 10 ones and trading them for 1 ten.

Name_____

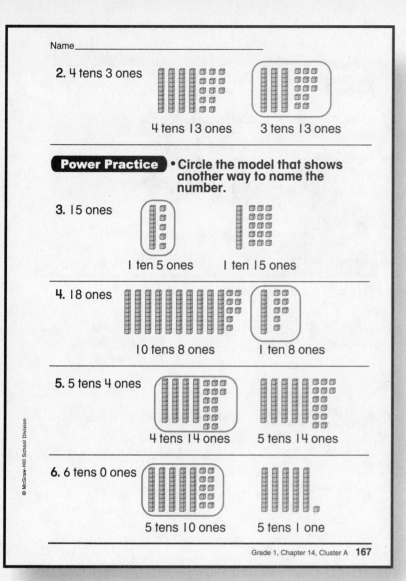

2. 4 tens 3 ones
   4 tens 13 ones   3 tens 13 ones

Power Practice • Circle the model that shows another way to name the number.

3. 15 ones
   1 ten 5 ones   1 ten 15 ones

4. 18 ones
   10 tens 8 ones   1 ten 8 ones

5. 5 tens 4 ones
   4 tens 14 ones   5 tens 14 ones

6. 6 tens 0 ones
   5 tens 10 ones   5 tens 1 one

USING THE LESSON

Try It
- Some students may benefit from having tens and ones models with which to show numbers.
- Do exercise 1 with students. Ask:
- *How many ones do you start with?* (24) *How many tens can you make from 24 ones?* (2 tens) *How many ones are left?* (4 ones) *Which choice is correct?* (2 tens 4 ones) Have students trace the circle.
- Have students complete exercise 2. Review their answers.

Power Practice
- Have students complete the practice items. Then review each answer.
- Ask: *In which exercises do you trade 1 ten for 10 ones?* (exercises 5-6) *In which exercises do you trade 10 ones for 1 ten?* (exercises 3-4)

Learn with Partners & Parents
Students can practice renaming by using dimes and pennies.
- Give the student a pile of pennies. Hold up a dime. Ask:
- *How many pennies equal 1 dime?* (10)
- Repeat with 2 dimes and 3 dimes. Then display 10 pennies. Ask:
- *How many dimes equal 10 pennies?* (1)
- Repeat with 20 pennies and 30 pennies.

WHAT IF THE STUDENT CAN'T

Complete the Power Practice
- Discuss each incorrect answer. Use these models: (exercise 3) *How many ones do you start with?* (15) *How many tens can you make from 15 ones?* (1 ten) *How many ones are left?* (5 ones) *Which choice is correct?* (1 ten 5 ones) (exercise 5) *How many tens and ones do you start with?* (5 tens 4 ones) *How many ones do you have if you trade 1 ten for 10 ones?* (14 ones) *How many tens are left?* (4 tens) *Which choice is correct?* (4 tens 14 ones) Have the student circle each correct answer.

USING THE LESSON

Lesson Goal
- Subtract facts to 20.

What the Student Needs to Know
- Recognize the meaning of subtraction.
- Recognize minus and equal signs.
- Subtract horizontally and vertically.

Getting Started
- On a felt board, show a group of 7 objects. Take 4 away. Have students say the subtraction sentence and solve for the difference. (7 − 4 = 3)
- Repeat with 6 objects, taking away 2; 12 objects, taking away 3; and 14 objects, taking away 8. (6 − 2 = 4, 12 − 3 = 9, 14 − 8 = 6))

What Can I Do?
- Read the question and the response. Then read and discuss the example. Ask:
- *How many flowers do you see?* (17) *How many flowers are taken away?* (7) *How can you find how many flowers are left?* (Count what is left; subtract.)
- *What number sentence can you write to subtract 5 flowers from 13 flowers?* (13 − 5 = 8) *Which number is the greatest number in a subtraction sentence?* (first)

Try It
- Ask: *How is the subtraction in exercise 1 different from the subtraction in exercise 2?* (One is written across; the other is written down.)
- Complete exercise 1 as a class. Say: *How many flowers do you see?* (7) *How many are subtracted?* (5) *Cross out 5. How many flowers are left?* (2)

Power Practice
- Have students complete the practice items. Then review each answer.

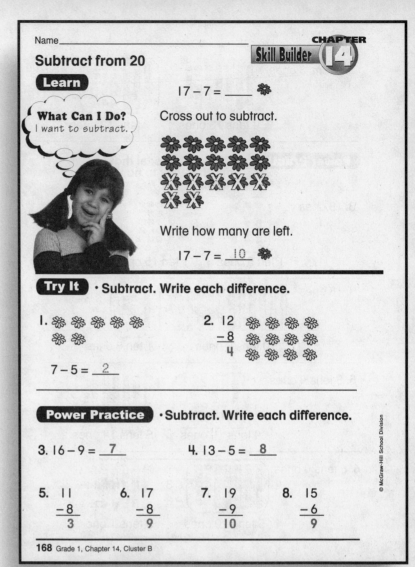

WHAT IF THE STUDENT CAN'T

Recognize the Meaning of Subtraction
- Have the student use the felt board and felt shapes to show these subtractions: 7 − 2, 5 − 4, 9 − 6.

Recognize Minus and Equal Signs
- Write number sentences with circles in place of the minus and equal signs. For example: 13 ◯ 9 ◯ 4. Have students supply the signs and read the sentences aloud.

Subtract Horizontally or Vertically
- Ask the student to rewrite exercises 3–4 as vertical addition problems and exercises 5–8 in horizontal form. Discuss how this affects the difference. (It doesn't.)

Complete the Power Practice
- Discuss each incorrect answer. Have the student model any fact he or she missed using counters.
- Have the student write each correct answer.

Name_____

Subtract: Counting Back

Skill Builder CHAPTER 14

Learn

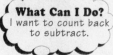

What Can I Do?
I want to count back to subtract.

11 – 3 = ___

Count back to subtract.

Count back 3: 11-10-9-8

11 – 3 = __8__

Try It
Count back to subtract. Write the difference.

1. 10 – 2 = __8__
2. 9 – 1 = __8__

Power Practice
Count back to subtract. Write the difference.

3. 7 – 1 = __6__
4. 11 – 2 = __9__
5. 10 – 3 = __7__
6. 12 – 3 = __9__

7. 6
   −2

    4

8. 12
   −2

   10

9. 8
   −3

    5

10. 11
    −1

    10

Grade 1, Chapter 14, Cluster B **169**

WHAT IF THE STUDENT CAN'T

Count Backward from 12
- Place 12 counters on the table. Ask the student to count them. Take away one at a time and have the student count backward to 0.

Read a Number Line
- Draw a 0-12 number line on the chalkboard. Have the student locate numbers as you say them aloud.
- Use a 0-12 number line. Ask: Which numbers are greater than 5? (6, 7, 8,. . .12) Which numbers are less than 7? (0, 1, 2,. . . 6)

Complete the Power Practice
- Discuss each incorrect answer..
- Have the student show you how to find the differences using the number line. Have him or her write each answer correctly.

USING THE LESSON

Lesson Goal
- Count back to subtract.

What the Student Needs to Know
- Count backward from 12.
- Read a number line.

Getting Started
- Have students count aloud from 0 to 5 and then back from 5 to 0. Then have them count from 0 to 12 and back from 12 to 0.

What Can I Do?
- Read the question and the response. Then read and discuss the example. Ask:
- *When you subtract 3 from 11, where do you start on the number line?* (at 11)
- *How many steps do you count back to subtract 3?* (3 steps)
- *Suppose you subtract 8 from 12. Where do you start?* (12) *How many steps do you count back?* (8)

Try It
- Students may use the number line to find the differences. Give students these steps to follow as you do exercise 1 together:
- *Find the first number on the number line. Put your finger there.*
- *To subtract 2, count back 2 spaces. Put your finger there. What is the difference?* (8) Have students trace the numbers.
- Have students complete exercise 2. Then check their answers.

Power Practice
- Tell students that they may use the number line if they wish.
- Have students complete the practice items. Then review each answer.

Grade 1, Chapter 14, Cluster B **169**

USING THE LESSON

Lesson Goal
- Subtract using doubles.

What the Student Needs to Know
- Add doubles.
- Recognize the inverse relationship between addition and subtraction.
- Recognize minus and equal signs.

Getting Started
- Give each pair of students 8 counters. Have the students arrange them in 2 rows of 4 counters.
- On the board, write 4 + 4 = 8.
- Have the students take away 1 row of counters. Write 8 − 4 = 4

What Can I Do?
- Read the question and the response. Then read and discuss the example. Ask:
- *How would knowing 5 + 5 help 10−5?* (If you know that 5 + 5 = 10, you know that you can reverse the numbers and write 10 − 5 = 5.)
- *What subtraction sentence would you write for 2 + 2 = 4?* (4 − 2 = 2); 6 + 6 = 12? (12 − 6 = 6)

Try It
- Review exercise 1 with the class before having students complete exercise 2 independently.

Power Practice
- Read the directions. Then have students complete the practice items. Review each answer.
- Challenge students to name the addition double that helps them solve each subtraction. (2 + 2, 1 + 1, 9 + 9, 6 + 6, 8 + 8, 10 + 10, 3 + 3, 5 + 5)

Name_____

Subtract Using Doubles

Learn

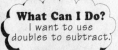

10 − 5 = ____

Think about doubles.

10 − 5 = ____
5 + 5 = 10

Write the difference.

10 − 5 = __5__

Try It • Subtract. Think about doubles. Write the difference.

1. 14 − 7 = __7__   2. 8 − 4 = __4__

Power Practice • Subtract. Write the difference.

3. 4 − 2 = __2__   4. 2 − 1 = __1__

5. 18 − 9 = __9__   6. 12 − 6 = __6__

7. 16   8. 20   9. 6   10. 10
   − 8   −10   − 3   − 5
   ───   ───   ───   ───
    8   10   3   5

WHAT IF THE STUDENT CAN'T

Add Doubles
- Put doubles from 1 + 1 through 9 + 9 on flash cards and have students quiz each other in pairs.
- Give the student an even number of counters. Have him or her make two equal rows of counters and write the addition double. Repeat with other even numbers.

Relate Addition to Subtraction
- Have the student use 8 counters to show that if 3 + 5 = 8, 8 − 5 = 3. Repeat with 2 + 6 = 8 and 8 − 6 = 2. Finish up with 1 + 7 = 8 and 8 − 7 = 1.

Recognize Minus and Equal Signs
- Write a variety of subtraction number sentences on the chalkboard and have the student read them aloud.

Complete the Power Practice
- Discuss each incorrect answer. Have the student model any fact he or she missed using parallel rows of counters, removing one row, and finding the difference. Then have the student write each correct answer.

Name_____

Subtract Related Facts

Skill Builder CHAPTER 14

Learn

What Can I Do?
I want to subtract related facts.

Subtract.

🌼🌼🌼🌼
🌼🌼🌼

8 − 5 = ____ 🌼🌼🌼
8 − 3 = ____ 🌼🌼🌼🌼🌼

Look for related facts.

8 − 5 = _3_ 🌼🌼🌼
8 − 3 = _5_ 🌼🌼🌼🌼🌼

Try It • Subtract.

1. 🌼🌼🌼🌼
   🌼🌼🌼

   7 − 3 = _4_
   7 − 4 = _3_

2.   9   9   🌼🌼🌼🌼🌼🌼🌼🌼🌼
   −3 −6
     6   3

Power Practice • Subtract.

3. 12 − 9 = _3_   4. 13 − 6 = _7_
   12 − 3 = _9_   13 − 7 = _6_

7. 8   8   6. 11   11   7. 15   15
   −6 −2   −4   −7   −8   −7
    2   6   7   4   7   8

© McGraw-Hill School Division

Grade 1, Chapter 14, Cluster B **171**

WHAT IF THE STUDENT CAN'T

Relate a Drawing to a Number Sentence

- On a felt board, show a row of 4 shapes and a row of 9 shapes. Have the student suggest two subtraction sentences that could be made using the display. (13 − 4 = 9, 13 − 9 = 4) Repeat with a row of 8 shapes and a row of 3 shapes. (11 − 8 = 3, 11 − 3 = 8)

Recall Subtraction Facts to 20

- Make up flash cards with subtraction facts to 20 and have pairs of students quiz each other.

Recognize Minus and Equal Signs

- Write several addition and subtraction number sentences and mix them up. Have the student circle all of the subtraction sentences and solve.

Complete the Power Practice

- Discuss each incorrect answer. Have the student use counters to model any exercise he or she missed. Then have the student write each answer.

USING THE LESSON

Lesson Goal
- Complete related subtraction facts.

What the Student Needs to Know
- Relate a drawing to a number sentence.
- Recall subtraction facts to 20.
- Recognize minus and equal signs.

Getting Started
- On a felt board, display a row of 5 shapes and a row of 7 shapes. Ask:
- *How many shapes are there in all?* (12)

Take away 5 shapes. Ask:
- *What subtraction sentence describes what I just did?* (12 − 5 = 7)
- Write the number sentence.
- Replace the 5 shapes. Take away 7 shapes. Ask:
- What number sentence describes what I just did? (12 − 7 = 5)
- Write the number sentence. Ask:
- *What do these two number sentences have in common?* (They are both subtraction sentences that use the same numbers.)

What Can I Do?
- Read the question and the response. Then read and discuss the example. Ask:
- *How would knowing 8 − 5 help you know 8 − 3?* (If you know that 8 − 5 = 3, you know that you can subtract the difference and write 8 − 3 = 5.)

Try It
- Review exercise 1 with the class before having students complete exercise 2 independently.

Power Practice
- Read the directions. Then have students complete the practice items. Review each answer.

Grade 1, Chapter 14, Cluster B **171**

CHALLENGE

Lesson Goal
- Recognize names for numbers.

Introducing the Challenge
- Ask: *How many ones are there in 3 tens and 5 ones?* (35 ones) *How can you make 3 tens 5 ones into 2 tens 15 ones?* (Trade 1 ten for 10 ones.)
- *Suppose you had 5 tens 2 ones. What would you have if you traded 1 ten for 10 ones?* (4 tens 12 ones)

Using the Challenge
- Read the directions aloud. Provide tens and ones models for those students who need them.
- If you wish, do the first problem as a group. Have students follow these steps:
- *How many tens are you given?* (2) *How many ones are you given?* (10) *Can you trade those ones?* (yes) *Now how many tens do you have?* (3) *What is that number?* (30) *Write it on the line. Then write it on the number line where it belongs.*
- After students have completed the page, discuss which problems they found easiest.

Name _____

Number Names

Write each number. Then write it in the correct box on the number line below.

1. 2 tens 10 ones = __30__   2. 1 ten 11 ones = __21__

3. 2 tens 13 ones = __33__   4. 1 ten 25 ones = __35__

5. 15 ones = __15__   6. 1 ten 16 ones = __26__

7. 2 tens 11 ones = __31__   8. 1 ten 9 ones = __19__

9. 23 ones = __23__   10. 1 ten 17 ones = __27__

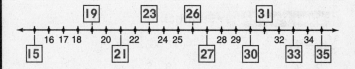

Name_____

Subtraction Match

Cut out the cards. Mix them up.
Put the cards face down in a pile. Play
with a partner.

To play: Each player takes 1 card from the
pile. The player with the greater difference
takes both cards. If the differences are the
same, the winner of the next round takes
all 4 cards.

20 – 10	14 – 17	14 – 9	9 – 7
18 – 9	12 – 5	13 – 9	11 – 9
16 – 7	12 – 6	6 – 2	7 – 6
16 – 8	14 – 8	10 – 7	10 – 9
18 – 10	10 – 5	11 – 8	12 – 2

Check that children are subtracting correctly as they play the game.

CHALLENGE

Lesson Goal
- Subtract facts to 20.

Introducing the Challenge
- On the chalkboard, write 5 – 2 and 4 – 3. Ask:
- *Which subtraction sentence has the greater difference?* (5 – 2)
- Now write 6 – 4 and 3 – 1. Ask:
- *Which subtraction sentence has the greater difference?* (The differences are the same.)

Using the Challenge
- Read the directions aloud. You may want to demonstrate turning up two cards, reading the facts, and having students determine which card has the greater difference.
- Have pairs of students play the game. Circulate to make sure they are following the rules.
- At the end of the game, the player who has more cards is the winner.